SOCIALIST AND POST-SOCIALIST URBANISMS

Critical Reflections from a Global H

Socialist cities have special qualities that endure in particular, subtle, and often under-theorized ways. This book engages with socialism on a global scale, as well as with a variety of socialist and post-socialist urbanisms and a range of ways in which globalization intersects with changes in socialist and post-socialist cities.

Offering a unique international comparative focus, the book's fourteen case studies from Asia, Europe, Latin America, and Africa are grouped under three main themes: housing experiences and life trajectories, planning and architecture, and governance and social order. Featuring contributors from a range of disciplinary backgrounds and research areas, *Socialist and Post-Socialist Urbanisms* brings together a collection of essays on cities that are often overlooked in mainstream urban studies.

LISA B.W. DRUMMOND is an associate professor in the Urban Studies Program in the Department of Social Science at York University.

DOUGLAS YOUNG is an associate professor in the Urban Studies Program in the Department of Social Science at York University.

Socialist and Post-Socialist Urbanisms

Critical Reflections from a Global Perspective

EDITED BY LISA B.W. DRUMMOND AND
DOUGLAS YOUNG

University of Toronto Press
Toronto Buffalo London

© University of Toronto Press 2020
Toronto Buffalo London
utorontopress.com
Printed in the U.S.A.

ISBN 978-1-4426-3283-7 (cloth)   ISBN 978-1-4426-3285-1 (EPUB)
ISBN 978-1-4426-3253-0 (paper) ISBN 978-1-4426-3284-4 (PDF)

Library and Archives Canada Cataloguing in Publication

Title: Socialist and post-socialist urbanisms: Critical reflections from a global
    perspective / edited by Lisa B.W. Drummond and Douglas Young.
Names: Drummond, Lisa B. W. (Lisa Barbara Welch), 1961– editor. | Young,
    Douglas, 1952– editor.
Description: Includes bibliographical references and index.
Identifiers: Canadiana (print) 20200154621 | Canadiana (ebook) 2020015463X |
    ISBN 9781442632837 (cloth) | ISBN 9781442632530 (paper) |
    ISBN 9781442632851 (EPUB) | ISBN 9781442632844 (PDF)
Subjects: LCSH: Sociology, Urban – Case studies. | LCSH: Cities and towns –
    Case studies. | LCSH: Socialism. | LCSH: Post-communism. | LCGFT:
    Case studies.
Classification: LCC HT151.S63 2020 | DDC 307.76–dc23

University of Toronto Press acknowledges the financial assistance to its
publishing program of the Canada Council for the Arts and the Ontario Arts
Council, an agency of the Government of Ontario.

Canada Council   Conseil des Arts
for the Arts   du Canada

ONTARIO ARTS COUNCIL
CONSEIL DES ARTS DE L'ONTARIO
an Ontario government agency
un organisme du gouvernement de l'Ontario

Funded by the   Financé par le
Government   gouvernement
of Canada   du Canada   Canada

Contents

Part Two – Planning and Architecture: Designing Socialist and Post-Socialist Urbanisms

Part Three – Governance and Social Order

Figures and Tables

Figures

Tables

Acknowledgments

This book would not have been possible without the contributions of many people. To the contributing authors we say thank you for your patience over the course of the *longue durée* of this project. The editorial staff at University of Toronto Press have been a pleasure to work with: acquisitions editor Douglas Hildebrandt in the beginning and Jodi Lewchuk as the book came closer to publication. We are grateful to the anonymous reviewers of the manuscript for their generous and helpful comments. Eric Clark gave freely of his time and wisdom and provided wonderful insights into the legacies of modernist urbanism. Ryan James provided valuable assistance with a review of existing scholarship on socialist and post-socialist cities.

Research we undertook with financial support provided by the Social Science and Humanities Research Council (SSHRC) of Canada (Standard Research Grant 410-2010-2617) for our project "Socialist Cities in the 21st Century: Modernist Legacies and Contemporary Policy-making" led eventually to this book project. Chapters 3 and 9 are based directly on the research we and our two wonderful post-doctoral fellows, Dr Markus Kip and Dr Nguyễn Thanh Bình, conducted in Hanoi and Berlin funded by that grant. The SSHRC grant also supported our participation at the 2013 Annual Meeting of the American Association of Geographers in Los Angeles; the papers presented in the two sessions we organized there galvanized the development of this collection.

Finally, together with the publisher we would like to thank the Faculty of Liberal Arts and Professional Studies at York University for the financial support it provided to this work in the form of two grants in aid of publishing.

Lisa B.W. Drummond and Douglas Young
Toronto, October 2019

SOCIALIST AND POST-SOCIALIST URBANISMS

Critical Reflections from a Global Perspective

Introduction: Socialist and Post-Socialist Urbanisms

LISA B.W. DRUMMOND AND DOUGLAS YOUNG

A city is a palimpsest of historical ideas about urbanism, urban idealism, and urban abandonment or deal-making, all of which have been materialized in the physical landscape or built environment. The early to mid-twentieth century was a period of dramatic urban transformation and expansion motivated by the application of theories of urban life and how urban spaces functioned. This was an era of science and idealism applied to urban form in a way that has rarely been glimpsed since. One scientific approach, socialist modernism, informed a range of urban projects, large and small, many of which are still extant and in use, though often not for the purposes or in the ways they were intended. Indeed, many cities are today grappling with questions of how to address the built remnants of ideas about urbanism that no longer pertain or have been wholeheartedly rejected. Yet despite their fall from contemporary fashion, social, political, and cultural concepts of an earlier era's urbanism may continue to be embodied in older urban residents who find themselves at odds with the policies of present-day urban regimes. Whether in bodies or in buildings, socialist urbanism endures in a variety of forms and locales. The contributions in this book examine how such reflections of the socialist past and its stubborn urban structures have been altered or adapted to, embraced or reviled, enveloped or obliterated, as societies in transformed or post-socialist cities live in their inherited environments. In this collection, we bring together case studies from around the socialist world to investigate and assess these debates, and to trace the fate of socialist urban projects in contemporary cities. Together, the chapters comprise a study of socialism as surviving in specifically existing modes and sites. For residents and leaders of current or once-socialist cities and of cities with substantial socialist urban projects, an investigation of how other such cities and urban spaces have been discussed and debated, and adapted or abandoned, may offer useful insights.

What Is Socialist Urbanism? / Capitalist and Socialist Modernisms

Socialist and modernist urbanism both originate in a negative reaction to the industrial capitalism that unfolded in the countries of western Europe in the late eighteenth and early nineteenth centuries. That reaction was to capitalism as a way of organizing economic activity and social relations, and to capitalist urbanization processes and the types of urban space they generated. Capitalist urbanism and everyday life was, for the urban majority, a cruel and unsustainable reality of abject poverty, dangerous working conditions, and derelict, overcrowded, and unhygienic living conditions. Eventually, that reality was widely acknowledged and attention was focused on addressing the urban crises in Europe's industrial cities. There is an obvious sympathy between socialist and modernist urbanism in their shared critique of capitalist urbanism and their conceptualization of alternate urban socio-spatial landscapes, but while many nineteenth- and twentieth-century critics of capitalism were both socialist and modernist, not all modernists were socialist and not all socialists were modernists.

Actually existing socialism has taken many forms, ranging from social democracy to repressive state socialism. Residents have experienced different varieties and intensities of socialist urbanism, which can be thought of as existing on a spectrum rather than having a concrete definition with set and specific characteristics. The common thread of the "social" that runs through them all, however, is threefold: a commitment to the modernist idea of universal progress, a desire to create a socially egalitarian society, and a heightened degree of state intervention in processes of city-building and urban governance.

In many respects, it is an effort of some dexterity to untangle capitalist-modernist from socialist urbanism, as key aspects of capitalist modernism and socialism are deeply intertwined. Both, for example, are accused of being undemocratic in that scientific expertise outweighs democratic decision-making in all spheres, even those of the everyday. Both require a radical break with past and traditional ways of life, including ideas about cities and urban life (everyday life and society in general). Both emphasize industrialization as a national economic strategy. Both prize the application of science/scientific thought to social life at all scales from the intimately everyday to the grandiose national. And for both, the city is a key location for effecting these radical breaks and scientific interventions in society and social organization, requiring large-scale interventions in the urban fabric.

The desired socialist society, however, is collective and socially undifferentiated, while the desired capitalist society is not necessarily so.

Socialist urbanists sought an urban form that would facilitate and, at the same time, reflect and reinforce the reshaping of capitalist social relations. For capitalist modernists, the reconceptualization of urban life was more design-focused and not necessarily linked to the overthrow of capitalist economy and society. Nevertheless, some characteristics of urban ideas generated by the crises of industrial capitalism were common to both socialists and capitalist-modernists.

Noteworthy was a focus on urban functionalism and a belief that urban activity – economic and social – could be scientifically studied and analysed by experts, and subsequently optimized with a particular urban form. Socialists and capitalists believed that all cities everywhere could be understood in terms of a shortlist of essential functions and that those functions would be facilitated by devoting distinct districts within a city to each function. Economic production would take place in designated industrial zones and business districts, and social reproduction would be supported in quiet residential zones. A range of transport infrastructure would connect the different functional zones. This functionalist approach to urbanism was crystallized in the Athens Charter, a document based on the discussions held in 1993 as part of CIAM IV (Congrès International d'Architecture Moderne; Le Corbusier was the author of the Charter, which he named after the site of the fourth CIAM Congress) (Jeanneret-Gris [Le Corbusier] 1973). Both capitalist and socialist urban planners invoked the Athens Charter although socialist planners soon developed their own specific principles of urban design, such as the GDR's Sixteen Principles (for a more detailed discussion, see Kip, Young, and Drummond 2015).

Socialist and capitalist urbanists in the early twentieth century confronted the past and rejected it. And both shared a disdain for pre-modern urbanism. In the case of socialist urbanism, this disdain incorporated a condemnation of capitalist city building. In practice, both evinced, to varying degrees, a desire to erase the legacies of the previous era and/or build from scratch on greenfield sites. An interesting example of the potential, but not necessary, intertwining of socialist and capitalist urbanisms in the twentieth century is Brasília, the new capital city developed in the late 1950s as a symbol of a postcolonial and modern Brazil. It was planned and designed by communists for a democratically elected federal government, and seen as a representation of Brazilian progress and modernity. Yet, as Richard Williams notes, "from 1964 on it became associated with a military regime that inadvertently had found in its epic spaces a representation of authority" (Williams 2009, 122). Brasília's modernism was thus reinterpreted and appropriated by an antidemocratic regime that seized power in a military coup.

Socialism, as a system of governance and as a source for ideas about how to "perfect" the city, has been a worldwide phenomenon. Every continent, with the possible exception of North America (and Antarctica), has housed at least one socialist government and at least one experiment in socialist urbanism. In Europe, the list of socialist states that intervened in urban projects or whole-scale urban development includes the Soviet Union, Poland, the German Democratic Republic, Hungary, Czechoslovakia, Yugoslavia, Bulgaria, Romania, Albania, Belarus, Poland, Ukraine, and Sweden (an exemplar of social democracy), and should, arguably, include British postwar urban interventions. In Africa, Benin, Congo, Ethiopia, Mozambique, Tanzania, Egypt, Libya, Sudan, Algeria, Cape Verde, the Seychelles, Madagascar, Angola, Somalia, and Guinea-Bissau have all experienced socialist states or socialist urban projects, as did the People's Democratic Republic of Yemen and Syria, as well as Iraq and Afghanistan in the Middle East. In Asia, China, Vietnam, Laos, Kampuchea (Cambodia), North Korea, Burma (now Myanmar), and Mongolia continue to be ruled by socialist or communist parties. And in Latin America and the Caribbean, Cuba, Nicaragua, Guyana, Venezuela, Chile, Peru, and Grenada were or are socialist states. Even in Australia, Canberra, the capital city, was planned according to Ebenezer Howard's "garden city" socialist principles of cooperative planning with a mix of social classes co-resident in neighbourhoods. In a number of additional countries socialism was or is noted as a national goal (in the state constitution, for example); these include the People's Republic of Bangladesh, the Republic of India, Nepal, Portugal, Sri Lanka, and Tanzania.

Notably absent from this list of countries that have had socialist governments and consequent socialist urbanisms are Canada and the United States of America. Both countries did develop social welfare states in the twentieth century but, while doing so, were resolutely determined to remain capitalist and anti-socialist. Perhaps the socialist impetus did not take hold in those countries in the twentieth century because neither had experienced the destruction caused by the First World War (as countries in Europe had), or the tumult of overthrowing colonial regimes (as was the case in many countries in Africa and Asia). Social welfare in Canada and the United States was developed in response to the Great Depression of the 1930s and as part of a Fordist social contract between labour and government. Clearly the Depression was an urgent social crisis in need of government response, but one that was to be addressed in Canada and the United States within capitalism and not by overthrowing it. For example, after the Second World War, the Canadian federal government launched a public housing program but it did so reluctantly and always with an eye to avoiding any negative impact on private sector landlords. Public housing from the 1940s until the 1970s was conceived of as residual housing intended for households with no other housing options (Dennis and Fish 1972; Sewell 1994).

Perhaps the earliest experiments in radically reimagining socio-spatial landscapes in response to the cities of the Industrial Revolution were those based on the ideas of nineteenth-century utopian socialists Robert Owen and Charles Fourier, which called for "a new moral world," "a new industrial world," and "a new amorous world." Dolores Hayden notes that "in the United States, Owen inspired about fifteen experiments in model community building beginning in the 1820s; Fourier inspired about thirty Associations, or Phalanxes, based on his ideas, beginning in the 1840s. (A smaller number of both types were conducted in Europe as well)" (Hayden 1982, 35).

Owenite and Fourierist model villages were built outside of cities; the earliest built urban example of a new socialist way of living might be the Familistère or Social Place at Guise in northern France, built by the industrialist (and Fourierist) Jean-Baptiste André Godin.

Beginning in 1859, Godin built workers' housing and community facilities adjacent to his iron foundries. A few decades later, he transferred ownership of the entire complex to the workers to operate as a cooperative (Hayden 1982; Benevolo 1967). The foundry continued to function into the 1960s (http://www.familistere.com/).

Other utopian socialist thinkers, such as those in the English Garden Cities movement in the late nineteenth century, addressed the challenge of reforming urban landscapes within the constraints of capitalism. It was not until after the Bolshevik Revolution in 1917 and the subsequent establishment of the Soviet Union that radically different socialist urbanisms were actually built. And in the rest of western Europe, it was the rupture of the First World War that allowed, after its conclusion in 1918, radical modernist urbanisms to be built in places such as the social democratic Weimar Republic.

Socialist Urbanism

After the Second World War, the socialist bloc expanded dramatically and extensively around the globe to Asia, Africa, and Latin America. In these new and in existing socialist states, planning was key state activity, often at every level, including planning of production and distribution, price setting, rationing, and housing allocation, but especially, for us, urban planning (Bernhardt 2005; Crowley and Reid 2002; DeHaan 2008, 2013; Fisher 1962; Forbes and Thrift 1987; French 1995; Hubacher

2000; May 2003; Rietdorf, Liebmann, and Schmigotzcki 2001; Smith 1996; Szelenyi 1982; Pióro, Savic, and Fisher 1965). Just as we noted that socialism existed on a spectrum, one that widened greatly with this expansion of the socialist bloc, the degree to which the socialist state assumed authority for urban planning varied. Where the prescriptive characteristics and strategies for capitalist modernist urbanism were detailed in the Athens Charter, socialist urbanism's foci and techniques, although heavily influenced by the Charter, were largely laid out by Soviet urban planning theorists and enshrined in policies such as the GDR's Sixteen Principles. Where the Athens Charter offered the categories of habitation, leisure, work, and transportation, Soviet-style planning, which theorized the importance of an entirely new social sphere, conceptualized the city as habitation, leisure, work, and culture.

The core unit in the Soviet urban system was the *mikroraion* or neighbourhood unit, which Smith (1996, 70) describes succinctly as conceptualized to provide "equality in housing conditions, local environmental quality and access to services" (see also Fisher 1962; Pióro et al. 1965). Spread evenly across the city, with aggregating densities for more specialist services (hospitals, in Smith's example), these would have provided all urban residents with equally accessible services that at the local level would have been within walking distance, and, more widely, easily reachable via plentiful public transit. In prosaic reality, however, new-built mikroraion did not completely replace existing (and differentiated) housing areas except where a city had been completely demolished in wartime or where a city was newly built to serve a particular and very large-scale industry with a specific geographic location (e.g., mines or nuclear reactors). In practice, not all areas of cities were provided with generous public transit, and not all housing estates, even if built in the mikroraion model, were equally well built or equipped with services of equal quality.

For socialist planners, city form was nevertheless a direct mode of pedagogical instruction towards achieving the socialist ideal society. At the same time, communist leaders in particular often viewed existing cities with ambivalence, even with suspicion of their capitalist origins and support for bourgeois lifestyles, and as sites containing those who had lived or aspired to bourgeois lifestyles. This was particularly the case for Asian communist states, where many of the cities existing at the moment of independence and instantiation of a communist regime were colonial products of an exploitative system against which the Communist Party, as an anti-colonial movement, had struggled violently. Nevertheless, cities were the key location of the transformation required to fully achieve socialism. The ideal citizen of this ideal society was the New Person, the New Socialist Person (Blum 2011; Chen 1969; Cheng 2009; Drummond 2004; duc Nhuân 1984; Mertin 2009; Samoff 1991; Simpson 2004; Wolfe 2005). This New Person had to be explicitly mobilized into existence through educational campaigns and the provision of a conducive built environment.

These urban interventions represent efforts in the work of transforming every aspect of everyday life through the relationships between work, home life (residence, consumption, recreation), and the national project (attaining socialist utopia). This national project was materialized in transformed spatial relationships within the city. In socialist societies most and often all formal employment was in effect state employment, as (virtually) all productive activities were nationalized. Residence became tied to workplace for many, especially industrial workers, as collective housing units were allocated to workplaces to be distributed to the work unit's workers. The sites of both production and reproduction were locations of mobilization to create the New Person in the national project of socialism.

The urban form of residential construction in Hanoi after independence under the Communist Party, for example, emphasized collective social life. Residences, especially in the new-built estates, were often assigned to state-run workplaces that then distributed the units to workers, tying domestic life firmly to working life, and thereby to state surveillance and intervention. Such state intervention also operated through other means, though workplace sanctions for domestic violations – exceeding the family size policy, for example – could be very effective. Households in the *khu tập thể*, or collective housing estates, were given units of sleeping and living space, but kitchens and toilets/bathing spaces were shared among groups of households. This was aimed both at encouraging more actively collective life and the use of collective services, such as canteens in the housing estates and in the workplace, and at economizing construction costs in times of desperate funding shortages. Despite socialist ideals of egalitarianism, however, housing was allocated according to "priority" (party status) and units varied in size, with larger units allocated to higher-ranking households (similarly, within the rationing system higher ranking households were given more generous allocations, including of rice and meat). In Vinh City, by contrast, as Christina Schwenkel details, the GDR-designed-and-constructed residential complex Quang Trung was made up of self-contained units that were accepted by the national Vietnamese authorities desperate to rebuild a war-flattened city, "despite Hanoi's anxieties about the moral contamination of its citizens by East German consumerism and individualism" (Schwenkel 2012, 451). In other words, collective society was to be created through communal living enforced by purpose-built housing; self-contained housing was tolerated in dire financial circumstances but was believed to undermine the project of social collectivization.

The transfer of socialist planning theories and concepts such as the mikroraion or neighbourhood unit was effected through a variety of means. Ideas circulated through scientific exchanges between the Soviet Union and other socialist/communist nations, through the loan of expertise between the Soviet Union or, in Christina Schwenkel's (2012) example, the GDR and Vietnam, and through academic training, for example as Vietnamese graduate students were sent to graduate or post-graduate programs in the USSR. William Logan (1995) points out in his analysis of Soviet planning influence in Hanoi over the thirty-five years between 1955 and 1990 that the Soviet Union posed not only as a model for Vietnam – an example of how to rehabilitate, restore, or rebuild urban areas damaged by war and an example of socialist urban form – but more directly, Vietnamese planners and architects were trained in Russia and in other Eastern European countries according to socialist theories of urbanism, with results visible in the design of landmark buildings, collective housing blocks and neighbourhoods, and industrial estates.

It was Soviet planners and architects who produced a master plan for Hanoi in 1973 (which called for the complete destruction of the city's Old Quarter and was, thankfully, never implemented). These circulations were not always or not only bilateral between the Soviet Union and other communist countries but also involved exchanges or loans of expertise among the wider community of socialist or non-aligned nations. Vietnamese experts, for example, went abroad to teach or advise in other socialist, often newly independent nations, as did the Hanoian chemistry professor sent to Guelma, Algeria in the 1980s whose experience is recounted by Susan Bayly (2007, 334–5; see also Bayly 2009).

Despite the persistence of socialist regimes currently in power, mainly but not exclusively in Asia, and while socialist urbanism continues to exist in the built environments of post-socialist cities, the heyday of socialist urban projects appears to have ended with the twentieth century. With its (near) demise in practice, the socialist city has flourished as a subject for academic study. Much of the English-language literature on the socialist city was triggered by the collapse of state socialist regimes in Central and Eastern Europe and the former Soviet Union in the late 1980s (Andrusz, Harloe, and Szelenyi 1996). Scholars addressed the question of whether or not there had been a distinctively socialist city and the extent to which socialist urban qualities had been universal or geo-temporally varied. Another key question addressed in the literature is whether it was urban spatial form itself, the methods used in creating it, or its use afterwards in everyday life that was distinctively socialist (May 2003); given, for example, that planners and architects in both capitalist and socialist societies subscribed to the same modernist Athens Charter notion of the functional city. Some scholars noted that there were many variations in European socialisms, and predicted a similar range of post-socialist urban trajectories (Marcuse 1996; Marcuse and Schumann 1992; Pickvance 2002).

Post-Socialist Cities

The rise of post-socialist cities corresponds in large part to the fall of the Berlin Wall and then the Soviet Union, but other non-European socialist regimes ended earlier or later and some continue. The term "post-socialist" seems most often used in the literature as a chronological marker, denoting the after of a before-the-Wall-fell. Where socialist states remain in power, commentators and even scholars often use "post-socialist" to describe the contemporary context of policy-making with strong neoliberal overtones, although China refers to "capitalism with Chinese characteristics" and Vietnam uses the language of "market socialism."

What is substantively "post-socialist" is a matter of some, though as yet surprisingly little, discussion. Significantly, such discussion is hampered by precisely the wide-ranging nature of socialisms-in-practice. Borén and Gentile (2007, 95) posit, with reference to central and eastern Europe, that the "post-socialist transformation [can be] understood as the economic, political, institutional and ideological changes associated with the discarding of 'communism' or 'state socialism' and the embracing of 'capitalism.'" And, as Kiril Stanilov (2007, 7) points out, in the aftermath of the demise of the Soviet Union, the cities of central and eastern Europe have undergone "radical transformations" at a rapid pace, and in myriad ways. The "post-socialist transformation" may indeed be dauntingly broad in concept, but undeniably its impact on urban spaces and urban lives has been concretely felt, though perhaps at a somewhat slower pace than changes in political and economic practices. Borén and Gentile (2007, 95) argue that certain socialist urban practices continue to exert influence; these "key significant socialist-era legacy aspects" include "central planning, land allocation, the second economy, defence considerations, and the implications of the ideological leadership of the communist parties." Despite scholarship's tendency to assume a post-socialist present, built environments are much more lethargic than policies, politicians, or individuals; a city takes a long time to change.

Analyses of post-socialist cities point to the institution of private ownership of property (especially housing; see Brown 1998; Hirt 2012; Owens 2014; Laszczkowski 2015) and of large sectors of the economy in general, planning which prioritizes development investment and in particular investment in land development (Golubchikov 2010; Badescu 2016; Cybriwsky 2016), and marketization or privatization of social services (e.g. health and education) and other urban services (e.g. water or transit) as being key shifts in urban practices with significant impact on residents' everyday and long-term experiences. Most often, scholars note post-socialism's urban effects in large-scale phenomena such as suburbanization and the development of gated communities (Špačková, Dvořáková, and Tobrmanová 2016; Fehervary 2011; Roose et al. 2013; Polanska 2010), in small-scale phenomena such as wealth-flaunting residences (Czegledy 1998; Cybriwsky 2016; Holleran 2014), and overall a range of varied and sometimes contradictory effects including downtown redevelopments (Scarpaci 2000), gentrification that displaces residents (Temelova et al. 2016), privatization of public spaces (Kalyukin et al. 2015), social polarization and marginalization (O'Neill 2011; Pedersen 2012; Polanska 2010), urban sprawl (suburbanization) or sometimes urban shrinkage (see Haase et al. 2016), and urban activism in a variety of forms.

In a key study of emergent post-socialist cities in Europe, Stanilov (2007, 6) characterizes the immediate preoccupations of post-socialist urban transformation as involving "housing privatization, property restitution, commercialization of the city centres, decentralization of housing and retail, growth of automobile ownership, etc." and notes via Bodnár (2001) that "privatization is the 'leitmotiv of post-socialist urban change.'" It is via privatization that urban Central and Eastern Europe began transitioning from "high-density, monocentric settlements, dominated by high-rise public housing and communal modes of transportation" to "sprawling, multi-modal metropolitan areas reaching extreme levels of privatization of housing, services, transportation, and public space" (Stanilov 2007, 7). In this rapid shift of urban orientation, Stanilov points to questions of sustainability. Where socialist urban planning proceeded without regard for the market, he notes that post-socialist cities' emphasis on the primacy of market forces is operating without regard to sustainability.

The core trend in post-socialist urbanism is the increased concentration of urban decision-making in the hands of non-state actors – the ascendance, in other words, of investment interests over comprehensive (or, often, even any) municipal planning (see also Ruoppila 2007). In this context, Stanilov (2007, 10) argues that regulations are ignored or rendered unenforceable because of a lack of state funding and public services are reduced. Investment interests drive piecemeal urban development, preferring the city core and the urban periphery where profits can mostly readily be realized. Informal developments, often including demolitions, are a different but equally piecemeal non-state intervention in a post-socialist context, which can also result in displacement or encroachment into green spaces (see Triantis and Vatavali 2016; Hirt 2008), as do large-scale state-backed and often internationally-funded developments. Post-socialist urbanism, in other words, offers both more – more choice in terms of things to buy and housing options – and much less, in terms of reduced public facilities and public space (Stanilov 2007, 11; see also Cybriwsky 2016 on access to the riverfront in Kyiv).

Place names and architecture also feature strongly in many cities experiencing a socialist to post-socialist transition. DeHaan (2008), in analysing the pre-Soviet, Soviet, and post-Soviet urban space of Nizhni Novgorod, argues that just as the city's socialist rulers did, the city's post-socialist authorities emphasized name changing – of streets, buildings, factories, and even cities (pre-Soviet Nizhni Novgorod became Gorky and after perestroika became Nizhni Novgorod again) – as important to conveying regime ideology. For many post-socialist regimes, as for post-socialist authorities in post-Soviet Russia, "rediscovering" the pre-socialist past is a key legitimizing strategy; that pre-socialist past is often manifest in the urban landscape, and in some cases, in urban morphology itself. DeHaan argues that post-socialist architecture is similarly implicated in communicating the city's post-socialist status. She shows that in post-Soviet Nizhni Novogord two styles of architecture have dominated in new constructions, what she calls "contextualized architecture" – buildings that reflect existing local pre-Soviet styles and ideas – and "generic modernism" – buildings that "would suit almost any industrial or post-industrial city" (280–2). Both suggest something about the desired or aspirational character of the post-socialist city (see also Stacul 2014; Young and Kaczmarek 2008; Golubchikov 2010; and the collection of papers in the 2013 special issue of *Nationalities Papers* guest edited by Diener and Hagen). Thus the city itself is a strategic element for state communication, especially in times of transition.

Reviewing the use of "post-socialist" in urban studies, Ferencuhová (2016, 519–21) argues that in scholarship on post-socialist cities she discerns two patterns of understanding. One sees the city as a passive entity upon which ebbs and flows, for example of capital and ideologies, work changes in the urban fabric, its governance, its people; the city is "shaped by external forces and political and economic changes" (520). The second asserts the agency of the city's residents in the changes to the urban landscape; she calls this "creative adaptation on the level of everyday life," which emphasizes the "active role of 'social agents' in interpretation and creation of on-going changes" (520). Ferencuhová notes that there has been little consensus over the "category of post-socialism" in urban studies (516) and she can offer only a few examples of scholarship (Grubbauer 2012; Czepczynski 2008; Stenning 2005; Stenning et al. 2010; Hirt 2012) that addresses "the specificity of the post-socialist urban experience" (Ferencuhová 2016, 516). It is in this latter effort that we see the contribution of this volume: place-specific insights into the fates of socialist spaces in contemporary cities.

This Volume

The collection of essays we have brought together here integrates case studies of socialist and post-socialist cities from around the world. In doing so, it offers comparisons of European, Asian, African, and Latin American socialism and post-socialism and juxtaposes studies that take different approaches to the study of socialist and post-socialist urbanism. We have organized these contributions under three broad themes: (1) housing experiences and life trajectories; (2) planning and architecture: designing socialist and post-socialist urbanisms; and (3) governance and social order. These themes encapsulate key aspects of socialist urbanism, and by extension are important features to consider in a transition to post-socialism – from the navigation of its everyday spaces to the grand ideas that motivated the design of those spaces and the policies regulating them.

Part One – Housing Experiences and Life Trajectories

Housing is one of the main areas in which socialist states undertake urban interventions. Most socialist states, many emerging in a postwar condition of considerable infrastructural damage, have faced severe housing shortages during part or all of the socialist period. Socialist states prioritized industrialization as a means to modernization, and thus often encouraged or induced urbanization (while severely, at least in Asian socialisms, curtailing freedom of movement). More than simply a necessity of urban life, socialist housing, as we described earlier, was expected to function as an important institution in the development of a socialist way of life. In cities where much of the housing stock had been damaged or was already severely crowded, accommodation in the newly constructed housing estates was often a privilege and highly desirable; it also materialized a set of pedagogic relationships between the family unit or the worker, the workplace, and the state. Housing construction, and construction of the "right type" of housing, was thus an important urban responsibility. The contributions in this section illuminate the human and everyday functions and interactions with housing in Asian and European socialist cities.

Christina Schwenkel, examining the contemporary condition of the flagship collective housing development in Vinh City, Vietnam, considers the repercussions of housing choice in different eras. When the Quang Trung estate, built with material and technical support from the GDR, opened its doors to residents in the early 1980s, being allocated an apartment in one of its buildings was considered a stroke of luck. But twenty to thirty years later, under "market socialism," the estate is derelict and residents have lost out in terms of land ownership. Now that the state confers land titles, it is vastly more desirable to own a plot of land than an apartment above ground, especially when that apartment is located in an estate that has seen little to no maintenance since construction. Lisa B.W. Drummond and Nguyễn Thanh Bình also examine the fate of socialist housing in Vietnam, seeing a similar trajectory in the desirability of collective housing estate units in Hanoi. In their chapter, Drummond and Nguyễn consider the everyday adaptations that residents made to their units and to the estates in general, and note the declining esteem in which these housing developments are held. Today the municipal government is unapologetic about its plans to eradicate the collective housing blocks and replace them with privately built and market-sold condominiums.

How residents navigate their housing options under a socialist regime is the focus of Thomas Borén and Michael Gentile's chapter, which chronicles the housing trajectory of one individual (and her family) in Leningrad. The details that emerge from their careful life-history (or rather housing-history) research with a citizen of the then-USSR demonstrate how this system was experienced in practice and show that although seemingly monolithic and immoveable, the restrictive state system could be negotiated and choice could be exercised, though with careful, highly specific, tactics. These engaging interviews with the resident also illuminate how residents understood the competing merits of what would otherwise seem to be virtually similar unit designs. In practice, even within patterns there are variations and residents were highly attuned to their resulting virtues or disadvantages.

Even in the social-democratic context of Sweden that Larsson investigates, housing was heavily invested with meaning and intention, and scrupulous attention was paid in the design of Vällingby suburb to create conditions as ideal as possible for the smooth functioning and nurturing of a healthy community. Larsson, a researcher at the Stockholm City Museum, details residents' everyday experiences of Vällingby. His account achieves a longitudinal study of the suburb where he revisits the addresses of participants in an original sociological study of Vällingby residents done in the 1950s. He documents changes in the residents and the residential experience over time, as well as changes in meaning of the space of Vällingby itself.

Part Two – Planning and Architecture: Designing Socialist and Post-Socialist Urbanisms

These chapters focus on the landscape of socialist urbanism, both in its contemporary moment and as a historical underpinning to contemporary urbanism. Socialist planners and architects, in their goal of perfecting the city – which they viewed as having failed under capitalism – undertook ambitious projects to redesign the urban centres of their cities and, in some cases, to design and plan entirely new cities (most of these new cities were built in the USSR). Architects sought ways to convey socialist ideas and ideals through building design. These chapters consider those projects and their aftermath in the post-socialist urban present. A number of them ask specifically how post-socialist societies inhabit and interact with socialist urban spaces.

Steven Logan examines 1960s housing developments in Prague, noting the convergence of prefabricated construction techniques and the scale of construction they permitted, alongside a planning emphasis on optimizing mobility through separating modes of transportation (pedestrian and automobile). He examines the conflict created in this effort to design a functionally appropriate socialist estate, as the exigencies of construction (the equipment needed to erect pre-fab buildings) overruled planning, design, and architectural decisions while the emphasis on separation of modes of mobility undermined the social goals of planning and design. He then considers how South City, Prague, is finding favour again among residents who appreciate the ideal of the "complete city" that motivated its original design.

Meanwhile, in Phnom Penh, Gabriel Fauveaud investigates the socialist spatialities underpinning this post-socialist city, considering how transition has affected the urban framework. This chapter focuses on the relationship between Phnom Penh's spatial transformation and its political and economic transformation, particularly the interaction between historical continuities and divergences in the urban landscape and the changes at various scales wrought by residents, developers, and local authorities.

A rare and valuable contribution focusing on African socialism, Jesse McClelland's chapter positions Addis Ababa in the context of African socialisms, examining the political transformations reflected in urban space. McClelland examines early socialist interventions that emphasized redistribution more than producing new spaces, as well as late and ineffectual (unimplemented) master planning efforts. In the contemporary context, urban space takes on increasing importance as the predominantly rural state experiences urbanization, and as efforts to industrialize and modernize lead to global urban connections. The chapter examines the grand urban visions of today's municipal authorities and their implications for Addis Ababa's urban residents.

In Europe again, Laura Visan offers us an insight into Romanian socialist urbanism, investigating the intended and actual fates of two urban sites designed by the Ceauşescu regime. One, the grandiose House of the Republic with its adjacent Victory of Socialism boulevard, was intended to create a new and socialist urban core for Bucharest. The other was a set of "alimentary centres" designed to deliver "scientific nutrition" despite the existing context of Romania's chronic food shortages; these monumental buildings became known as "hunger circuses." With the abrupt end of the Ceauşescu regime, these unfinished sites were quickly abandoned. Visan follows the debates over how these sites were to be repurposed or overlooked in the post-socialist urban landscape.

Markus Kip and Douglas Young similarly focus on contemporary debates over sites of socialist urban heritage, in this case Alexanderplatz, the iconic square at the spiritual centre of East Berlin and the GDR. Kip and Young trace the debates, following German unification in 1990, about the fate of socialist and modernist Alexanderplatz, noting the historical back and forth between socialist and modernist urbanism. Ultimately, they conclude that those who argue in favour of preserving Alexanderplatz in its currently existing form are in fact anti-modern in spirit. The authors suggest that the debate over Alexanderplatz should not simply be one of economic growth and development versus architectural heritage preservation. Instead, the social uses of the space should be of paramount concern.

Part Three – Governance and Social Order

The chapters in this section consider the life and afterlife of socialist governance – the lingering influence of socialist systems that continue to affect contemporary post-socialist or "market socialist" urban life in myriad ways. These chapters ask how past socialist governance has structured an urban present. Where can we see continuities, perhaps under different guises and serving different interests, and where do we see distinctions or new paths and systems?

Governance through mobilization is the topic of Carolyn Cartier's chapter, which examines the National Civilized City campaign as a technique of governance in the context of China's contemporary socialism. The National Civilized City program reveals the top-down and highly bureaucratized nature of socialist urbanism in China, though at the same time a technique so focused on self-regulation is also highly neoliberal. Cartier situates this civilizing program in the history of the "capitalism with Chinese characteristics" that has emerged out of "socialism with Chinese characteristics." She highlights how this program, which singles out urban successes with the designation "National Civilized City," fits neatly within a Chinese tradition of modelling for emulation, which predates but dovetails neatly with the global socialist model-making tradition and emulation campaigns for achieving socialist goals. In this chapter, Cartier shows how the campaign works to encourage self-regulation on the part of municipal authorities and, simultaneously, to allow discipline by the central state in the bestowal, withholding, or withdrawal of the coveted title.

Two contributions in this section further the volume's goal of illuminating under-studied aspects or sites of socialist urbanism. Marcela Mele and Andrew Jonas's chapter focuses on suburban governance in Tirana, Albania. Here, the authors consider how the transition from socialism has structured Tirana's urban periphery in ways that converge, but also diverge, from global trends in suburban governance. Mele and Jonas argue that Albania's experience of anti-urban communism positioned the country's cities, and specifically their recently rural suburbs, to experience the transition to capitalism through phenomena ranging from informality to struggles over political citizenship. Informality has been particularly important to modes of contemporary suburban governance because it figures heavily in the livelihood strategies by which suburbanites have staked their claims to land, services, and a political voice.

Laura Shillington similarly offers insight into an understudied socialist context, that of Nicaragua. Presenting a rare political ecology-based examination of Managua's socio-natural relations, Shillington examines how the contemporary form of these relations have been shaped through discourses of socialism and modernism. Shillington investigates how lakes and earthquakes, which she argues are definitive of Managuan urbanism, have been discursively transformed under Sandinistan socialism, neo-liberalism, and the contemporary neo-Sandinista regime from sites of revolutionary potential to sites of privatization and global concerns over sustainable development and to sites of neoliberal socialist nostalgia.

Focusing on young people in Ho Chi Minh City, Emmanuelle Peyvel and Marie Gibert analyse how their young interviewees engage in lifestyle activities under conditions of market socialism. Peyvel and Gibert argue that leisure practices have transformed from pursuits that encouraged the embodiment of socialist ideals to activities emphasizing the private and the intimate in contemporary society. The chapter offers a useful case study of the ways in which young professionals, in this case, negotiate the legacies of socialist urbanism while aspiring to global lifestyles and social mobility. The authors consider how contemporary leisure practices may reflect both socialist and neoliberal norms.

The final contribution again offers insight into a new way of thinking about the lingering impact of socialist governance. Maps are an often-neglected resource for insight into political, social, and personal ideas of the city. This chapter, like many others in the volume, analyses a city about which we hear and know little in the English-language literature, Khujand in Tajikistan, on the periphery of the Soviet Union. In this chapter, Wladimir Sgibnev analyses a variety of maps, from official to mental, from a variety of eras, pre-Soviet to contemporary. His analysis of these maps considers how residents think about the urban space they inhabit, and the legacies of socialism and the post-socialist transition for shaping those understandings.

Individual modes of orientation, way-finding, and landmarking are important for knowing the city, especially where a socialist regime restricts access to maps. But even individual modes of urban mapping are influenced by official cartographic conceptualizations of the city. As Sgibnev demonstrates, where these ignore particular aspects of the city – the old town, the nonmonumental everyday city – that deliberate erasure is internalized and reflected in the residents' own maps and a particular sampling of the range of socialist and post-socialist urbanisms emerge. In all the chapters, however, we hear the voices and see the purposive traces of those who have lived and are living these urban concepts as residents who must negotiate these material forms.

The Conclusion to this volume offers an analysis of the contributions presented here, by posing four dimensions of space to help unpack the key question of what is socialist about an urban space. We review how the chapters offer insights into these four dimensions of space and contribute to understanding socialist spatiality. In contemporary cities, are these socialist spaces what DeHaan (2008, 288–9) in an archaeological metaphor calls a "contaminant," poking up from an earlier layer to contaminate the post-socialist present? Are they mere remnants, disregarded and uncelebrated in the urban landscape though they may still be in use? Or are they valued monuments to past ideas perhaps no longer dominant but acknowledged as part of the spectrum of human ideas about how to live? The chapters in this volume present a range of experiences and approaches to socialist spaces in post-socialist cities which illuminate the very different ways in which these spaces exist in the urban present.

REFERENCES

Andrusz, Gregory, Michael Harloe, and Ivan Szelenyi, eds. 1996. *Cities after Socialism: Urban and Regional Change and Conflict in Post-Socialist Societies*. Oxford, UK: Blackwell.

Badescu, Gruia. 2016. "(Post) Colonial Encounters in the Postsocialist City: Reshaping Urban Space in Sarajevo." *Geografiska Annaler: Series B, Human Geography* 98, no. 4: 321–9.

Bayly, Susan. 2007. *Asian Voices in a Post-Colonial Age: Vietnam, India and Beyond*. Cambridge, UK: Cambridge University Press.

– 2009. "Vietnamese Narratives of Tradition, Exchange and Friendship in the Worlds of the Global Socialist Ecumene." In *Enduring Socialism: Explorations of Revolution and Transformation, Restoration and Continuation*, edited by Harry G. West and Parvathi Raman, 125–47. New York, NY: Berghan Books.

Benevolo, Leonardo. 1967. *The Origins of Modern Town Planning*. Cambridge, MA: MIT Press.

Bernhardt, Christoph. 2005. "Planning Urbanization and Urban Growth in the Socialist period: The Case of East German New Towns, 1945–1989." *Journal of Urban History* 32, no. 1: 104–19. https://doi.org/10.1177/0096144205279201

Blum, Denise F. 2011. *Cuban Youth and Revolutionary Values: Educating the New Socialist Citizen*. Austin: University of Texas Press.

Bodnár, Judit. 2001. *Fin de Millénaire Budapest: Metamorphoses of Urban Life*. Minneapolis: University of Minnesota Press.

Borén, Thomas and Michael Gentile. 2007. "Metropolitan Processes in Post-Communist States: An Introduction." *Geografiska Annaler: Series B, Human Geography* 89, no. 2: 95–110.

Brown, Andrew J. 1998. "Taking Shelter: The Art of Keeping a Roof Overhead in Post-Soviet Almaty." *Central Asian Survey* 17, no. 4: 613–28.

Chen, Theodore Hsi-En. 1969. "The New Socialist Man." *Comparative Education Review* 13, no. 1: 88–95. https://doi.org/10.1086/445389

Cheng, Yinghong. 2009. *Creating the New Man: From Enlightenment Ideals to Socialist Realities*. Honolulu: University of Hawaii Press.

Crowley, David, and Reid, Susan E., eds. 2002. *Socialist Spaces: Sites of Everyday Life in the Eastern Bloc*. Oxford, UK: Berg Publishers.

Cybriwsky, Roman Adrian. 2016. "Whose City? Kyiv and Its River after Socialism." *Geografiska Annaler: Series B, Human Geography* 98, no. 4: 367–79.

Czegledy, Andre. 1998. "Villas of Wealth: A Historic Perspective on New Residences in Post-Socialist Hungary." *City and Society* 10, no. 1: 245–68.

Czepczyński, Mariusz. 2008. *Cultural Landscapes of Post-Socialist Cities: Representation of Powers and Needs*. Farnham, UK: Ashgate.

DeHaan, Heather D. 2008. "Finding the Soviet in Post-Soviet Space: An Excavation of the Post-Soviet City of Nizhnii Novgorod." In *What is Soviet Now?*, edited by Thomas Lahusen and Peter Solomon, 277–305. London, UK: LIT Verlag.

– 2013. *Stalinist City Planning: Professionals, Performance, and Power*. Toronto, ON: University of Toronto Press.

Dennis, Michael, and Fish, Susan. 1972. *Programs in Search of a Policy: Low Income Housing in Canada*. Toronto, ON: Hakkert.

Diener, Alexander C., and Joshua Hagen. 2013. "From Socialist to Post-Socialist Cities: Narrating the Nation through Urban Space." *Nationalities Papers* 41, no. 4: 487–514.

Drummond, Lisa. 2004. "The Modern Vietnamese Woman: Socialization and Womens' Magazines." In *Gender Practices in Contemporary Vietnam*, edited by Lisa Drummond and Helle Rydstrøm, 158–78. Copenhagen, Denmark: NIAS Press.

duc Nhuân, Nguyên. 1984. "Do the Urban and Regional Management Policies of Socialist Vietnam Reflect the Patterns of the Ancient Mandarin Bureaucracy?" *International Journal of Urban and Rural Research* 8, no. 1: 73–89. https://doi.org/10.1111/j.1468-2427.1984.tb00414.x

Fehervary, Krisztina. 2011. "The Materiality of the New Family House in Hungary: Postsocialist Fad or Middle-Class Ideal?" *City and Society* 23, no. 1: 18–41.

Ferencuhová, Slavomira. 2016. "Explicit Definitions and Implicit Assumptions about Post-Socialist Cities in Academic Writings." *Geography Compass* 10/12: 514–24.

Fisher, Jack C. 1962. "Planning the City of Socialist Man." *Journal of the American Institute of Planners* 28, no. 4: 251–65. https://doi.org/10.1080/01944366208979451

Forbes, Dean K., and Nigel Thrift. 1987. *The Socialist Third World: Urban Development and Territorial Planning*. New York, NY: Basil Blackwell.

French, R. Antony. 1995. *Plans, Pragmatism and People: The Legacy of Soviet Planning for Today's Cities*. London, UK: UCL Press.

Golubchikov, Oleg. 2010. "World-City-Entrepreneurialism: Globalist Imaginaries, Neoliberal Geographies, and the Production of New St Petersburg." *Environment and Planning A* 42, no. 3: 626–43.

Grubbauer, Monika. 2012. "Toward a More Comprehensive Notion of Urban Change: Linking Post-Socialist Urbanism and Urban Theory." In *Chasing Warsaw: Socio-material Dynamics of Urban Change since 1990*, edited by Monika Grubbauer and Joanna Kusiak, 35–60. Frankfurt, Germany: Campus Verlag.

Haase, Annegret, Dieter Rink, and Katrin Grossmann. 2016. "Shrinking Cities in Post-Socialist Europe: What Can We Learn from Their Analysis for Theory Building Today?" *Geografiska Annaler: Series B, Human Geography* 98, no. 4: 305–19.

Hayden, Dolores. 1982. *The Grand Domestic Revolution: A History of Feminist Designs for American Homes, Neighbourhoods, and Cities*. Cambridge, MA: MIT Press.

Hirt, Sonia. 2008. "Landscapes of Postmodernity: Changes in the Built Fabric of Belgrade and Sofia since the End of Socialism." *Urban Geography* 29, no. 8: 785–810.

– 2012. *Iron Curtains: Gates, Suburbs and Privatization of Space in the Post-Socialist City*. Hoboken, NJ: John Wiley & Sons.

Holleran, Max. 2014. "*Mafia Baroque*: Post-Socialist Architecture and Urban Planning in Bulgaria." *British Journal of Sociology* 65, no. 1: 21–42.

Hubacher, Simon. 2000. "Berlin-Marzahn: The Would-Be Town." In *City of Architecture/Architecture of the City*, edited by Thorsten Scheer, Josef Paul Kleihues, and Paul Kahlfeldt, 349–60. Berlin, Germany: Nicolai.

Jeanneret-Gris, Charles Édouard [Le Corbusier]. 1973. *The Athens Charter*. New York, NY: Grossman.

Kalyukin, Alexander, Thomas Borén, and Andrew Byerley. 2015. "The Second Generation of Post-Socialist Change: Gorky Park and Public Space in Moscow." *Urban Geography* 36, no. 5: 674–95.

Kip, Markus, Douglas Young, and Lisa B.W. Drummond. 2015. "Socialist Modernism at Alexanderplatz." *Europa Regional* 22, no. 1–2: 13–26.

Laszczkowski, Mateusz. 2015. "Scraps, Neighbors, and Committees: Material Things, Place-Making, and the State in an Astana Apartment Block." *City & Society* 27, no. 2: 136–59.

Logan, William. 1995. "Russians on the Red River: The Soviet Impact on Hanoi's Townscape 1955–1990." *Europe-Asia Studies* 47, no. 3: 443–68. https://doi.org/10.1080/09668139508412266

Marcuse, Peter. 1996. "Privatisation and Its Discontents: Property Rights in Land and Housing in the Transition in Eastern Europe." In *Cities after Socialism: Urban and Regional Change and Conflict in Post-Socialist Societies*, edited by Gregory Andrusz, Michael Harloe, and Ivan Szelenyi, 119–91. Oxford, UK: Blackwell.

Marcuse, Peter, and Wolfgang Schumann. 1992. "Housing in the Colours of the GDR." In *The Reform of Housing in Eastern Europe and the Soviet Union*, edited by Bengt Turner, Jósef Hedegüs, and Iván Tosics, 74–144. London, UK: Routledge.

May, Ruth. 2003. "Planned City Stalinstadt: A Manifesto of the Early German Democratic Republic." *Planning Perspectives* 18, no. 1: 47–78.

Mertin, Evelyn. 2009. "Presenting Heroes: Athletes as Role Models for the New Soviet Person." *The International Journal of the History of Sport* 26, no. 4: 469–83.

O'Neill, Bruce. 2011. "Cast Aside: Boredom, Downward Mobility, and Homelessness in Post-Communist Bucharest." *Cultural Anthropology* 29, no. 1: 8–31.

Owens, Geoffrey Ross. 2014. "From Collective Villages to Private Ownership: *Ujamaa, Tamaa,* and the Postsocialist Transformation of Peri-Urban Dar es Salaam, 1970–1990." *Journal of Anthropological Research* 70: 207–31.

Pedersen, Morten Axel. 2012. "A Day in the Cadillac: The Work of Hope in Urban Mongolia." *Social Analysis* 56, no. 2: 136–51.

Pickvance, Chris. 2002. "State Socialism, Post-Socialism and Their Urban Patterns: Theorizing the Central and Eastern European Experience." In *Understanding the City: Contemporary and Future Perspectives*, edited by John Eade and Christopher Mele, 183–203. Oxford, UK: Blackwell.

Pióro, Zygmunt, Milos Savic, and Jack Fisher. 1965. "Socialist City Planning: A Reexamination." *Comments* 31, no. 1: 31–42.

Polanska, Dominika. 2010. "Gated Communities and the Construction of Social Class Markers in Postsocialist Societies: The Case of Poland." *Space and Culture* 13, no. 4: 421–35.

Rietdorf, Werner, Heike Liebmann, and Britta Schmigotzcki, eds. 2001. *Further Development of Large Housing Estates in Central and Eastern Europe as Constituent Elements in a Balanced, Sustainable Settlement Structure and Urban Development.* Erkner, Germany: Institute for Regional Development and Structural Planning.

Roose, Antti, Ain Kull, Martin Gauk, and Taivo Tali. 2013. "Land Use Policy Shocks in the Post-Communist Urban Fringe: A Case Study of Estonia." *Land Use Policy* 30, no. 1: 76–83.

Ruoppila, Sampo. 2007. "Establishing a Market-Orientated Urban Planning System after State Socialism: The Case of Tallinn." *European Planning Studies* 15, no. 3: 405–26.

Samoff, Joel. 1991. "Socialist Education?" *Comparative Education Review* 35, no. 1: 1–22.

Scarpaci, Joseph L. 2000. "Reshaping Habana Viera: Revitalization, Historic Preservation, and Restructuring in the Socialist City." *Urban Geography* 21, no. 8: 724–44.

Schwenkel, Christina. 2012. "Civilizing the City: Socialist Ruins and Urban Renewal in Central Vietnam." *Positions* 20, no. 2: 437–70.

Sewell, John. 1994. *Houses and Homes: Housing for Canadians*. Toronto, ON: James Lorimer.

Simpson, Pat. 2004. "Parading Myths: Imaging New Soviet Woman on Fizkul' turnik's Day, July 1944." *The Russian Review* 63, no. 2: 187–211.

Smith, David M. 1996. "The Socialist City." In *Cities after Socialism: Urban and Regional Change and Conflict in Post-Socialist Societies*, edited by Gregory Andrusz, Michael Harloe, and Ivan Szelenyi, 70–99. Oxford, UK: Blackwell.

Špačková, Petra, Nina Dvořáková, and Martina Tobrmanová. 2016. "Residential Satisfaction and Intention to Move: The Case of Prague's New Suburbanites." *Geografiska Annaler: Series B, Human Geography* 98, no. 4: 331–48.

Stacul, Jaro. 2014. "The Production of 'Local Culture' in Post-Socialist Poland." *Antrhopological Journal of European Cultures* 23, no. 1: 21–39.

Stanilov, Kiril. 2007. "Taking Stock of Post-Socialist Urban Development: A Recapitulation." In *The Post-Socialist City*, edited by Kiril Stanilov, 3–17. Dordrecht, The Netherlands: Springer.

Stenning, Alison. 2005. "Post-Socialism and the Changing Geographies of the Everyday in Poland." *Transactions of the Institute of British Geographers* 30, no. 1: 113–27.

Stenning, Alison, Adrian Smith, Alena Rochovská, and Dariusz Świątek. 2010. *Domesticating Neo-Liberalism: Spaces of Economic Practice and Social Reproduction in Post-Socialist Cities*. London, UK: Wiley Blackwell.

Szelenyi, Ivan. 1982. "Inequalities and Social Policy under State Socialism." *International Journal of Urban and Regional Research* 6, no. 1: 121–7.

Temelova, Jana, Jakub Novak, Anneli Kharkiv, and Tiit Tammaru. 2016. "Neighbourhood Trajectories in the Inner Cities of Prague and Tallinn: What Affects the Speed of Social and Demographic Change?" *Geografiska Annaler: Series B, Human Geography* 98, no. 4: 349–66.

Triantis, Loukas, and Fereniki Vatavali. 2016. "Informality and Land Development in Albania: Land Reforms and Socioeconomic Dynamics in a Coastal Settlement." *Geografiska Annaler: Series B, Human Geography* 98, no. 4: 289–303.

Williams, Richard J. 2009. *Brazil (Modern Architectures in History)*. London, UK: Reaktion Books.

Wolfe, Thomas C. 2005. *Governing Soviet Journalism: The Press and the Socialist Person after Stalin*. Bloomington: Indiana University Press.

Young, Craig, and Sylvia Kaczmarek. 2008. "The Socialist Past and Postsocialist Urban Identity in Central and Eastern Europe." *European Urban & Regional Studies* 15, no. 1: 53–70.

PART ONE

Housing Experiences and Life Trajectories

1 From Socialist Moderns to the New Urban Poor: Gender and the Housing Question in Vinh City, Vietnam

CHRISTINA SCHWENKEL

Upon arrival in the industrial city of Vinh in north central Vietnam, a provincial capital with half a million inhabitants, one comes across rows of crumbling housing blocks built by East Germany (GDR) as a "solidarity gift" to Vietnam after the end of the US-led air war in 1973. With twenty-two five-storey buildings and attenuating facilities spread across the city centre that housed more than 9,000 workers and civil servants, the sprawling housing estate, known as *khu chung cư Quang Trung*, or Quang Trung for short, showcased East Germany's technical capacities on the one hand, and Vietnamese aspirations to socialist modernity on the other.[1] From the wide boulevards that surround the settlement, the blocks appear uniform in style, like socialist cities in Eastern Europe with their trademark panel housing for the masses (whereas these were made of brick), but with a keener eye, from the narrow lanes that cut across the complex, one can discern subtle variations in aesthetic style and technical design, as well as different degrees of decline of the individual buildings. Each block represents a specific history of urban design, allocation, settlement, and use of built space in accordance with socialist housing policy – including its remaking and repurposing over the past forty years.

Indisputably, block C8, along the northern edge of the complex that faces a busy street corner and adjacent park, stands out as the building that has fallen into the most disrepair (see figure 1.1). According to local housing authorities, the physical quality of C8 is assessed at less than 40 per cent of its original structural capacity, generating fear among residents of its impending collapse. C8's notorious reputation as dangerous ruins has garnered considerable attention in the press, as shards of the building have literally fallen off and injured inhabitants. Because government officials have taken no direct action to ameliorate such risks to the population, block C8 – allocated in 1981 to preferential

1.1 Block C8 in Quang Trung, notorious in Vinh City for its dilapidated condition, 2011. Photo by Christina Schwenkel.

female workers at the state pharmaceutical company – has become emblematic of the inadequacies of urban governance and housing policies in the post-reform era of "market socialism." Accordingly, the women of C8 – once heroic subjects of history and now objects of public pity – have emerged as the new figure of the urban disenfranchised. As I argue in this chapter, this shift in subject positioning marks the women's symbolic reversal in status from socialist moderns to the new urban poor.

Changes to status and subjectivity have been noted elsewhere in scholarship on model socialist communities that experienced rapid transformation and deindustrialization following the adoption of economic and political reforms.[2] Here I bring attention to the gendered dynamics of such shifts, a perspective largely neglected in the literature, as the golden age of the worker has been replaced with that of the innovative entrepreneur. In Vinh City, also celebrated as a model, planned socialist city (in its early years after reconstruction), such changes have been linked to – but as I show, are not entirely the result of – the slow and steady move towards a market-oriented urbanism with the onset of Đổi mới economic reforms in 1986. Crucial to the women in block C8 (and residents in social housing more generally), these reforms entailed the dismantling of a system of centralized distribution of goods and services, along with other redistributive policies, including the provision of housing through the workplace, which served as the basic urban social unit under socialism.[3] As in other cities across Vietnam,[4] the privatization of housing, and the real estate booms and busts that followed, increased socio-economic inequality and its spatialization across the city,[5] although here I argue that such inequality cannot be fully disentangled from practices of housing distribution under socialism. Rather, the contemporary gender disparities that I discuss in this chapter are rooted in differential policies of housing allocation that existed prior to the onset of market reforms.

Historically, the relationship between cities and socialism was ambiguous and erratic, differing across time, space, and sphere of political influence. In the Soviet bloc, for instance, state modernization projects focused on urban construction in the aftermath of the Second World War, while the anti-urban stance of Maoists served to depopulate, and in some radical cases empty, entire cities.[6] The "forced demodernization" of cities, as Stephen Graham (2006) has observed of twentieth-century air wars, levelled urban landscapes on which new industrial centres could be built according to planning ideologies that emphasized city-making as a transformative, if not utopian, social mission. Urban planning, including the construction of housing, stood at the centre of governing strategies for regulating populations and producing a new generation of socialist men and women (Collier 2011). Ensuring equal access to urban infrastructure was to reflect both the ideological values and the lived experience of a prosperous and egalitarian society. As scholars of socialism have argued (e.g., Zavisca 2012; Fehérváry 2013; Harris 2013), housing, in particular, embodied the hopeful promise of socialism, with the provision of separate family apartments serving as an instrumental means to produce a population trained in the socialist arts of modern living. GDR planners in Vietnam likewise embraced this ideology in their ambitious reconstruction of Vinh City, seeing individual, self-contained units as a path to urban modernity and a tool to socially engineer a new humanity and reproduce labour power through the nuclear family. And yet, similar to the case of the Soviet Union and other countries, for certain members of the population, especially for residents in block C8, the vision of the separate apartment as the beacon of modern socialist living rarely materialized, due to postwar exigencies and the critical demand for housing that exceeded available supply.

In what follows, I examine how efforts to resolve the postwar housing crisis at a state pharmaceutical company backfired under market reforms, putting at great bodily risk and economic disadvantage the very female workers that revolutionary policy was designed to protect. These now-retired workers are disproportionately women because of a method of allocation that pegged housing distribution (and food rations) to official rank and a calculated point system according to people's need and contribution to nation-building. Although allocation preference was typically given to women with children with the intent to "stabilize" their lives (Hoang 1999, 90), housing options in Vinh City– land, collective housing, or an apartment in Quang Trung – were not equally ranked, valued, or distributed. Because of the gendered division of labour in the workplace, in the immediate postwar years men were more often allocated land than women, who were given apartments in Quang Trung. Such accommodation was considered by many at the time to be convenient and desirable, especially for female-headed households. These perceptions began to change, however, with the onset of Đổi mới. I compare how two groups of retirees from the pharmaceutical company – recipients of land and recipients of apartments – have fared under privatization policies that allow for the sale of land and property. Scholars of gender have argued that northern Vietnamese women, in particular, have been adversely affected by market reforms, losing many of the gains made by the revolution (see, especially, Werner 2009). For retired pharmaceutical workers, my findings confirm this observation: the increasing wealth of predominantly male land users contrasts sharply with the deepening poverty of female apartment owners.[7] Similarly, feminist scholars of Eastern Europe and the Soviet Union, such as Susan Gal and Gail Kligman (2000) have long noted the role that gender plays in uneven geographies of postsocialist development. In Vinh City, the precarious location of women in the contemporary housing economy (owing to a history of hierarchical distribution across gender) reaffirms their argument that privatization after the collapse of socialism has shaped and been shaped by gendered practices, discourses, and assumptions about biological difference that have affected men and women in profoundly different ways (Gal and Kligman 2000, 3).

Evacuation: Collective Housing in Wartime

During the US air war against northern Vietnam (1964–73), the violence of relentless bombing forced entire villages to evacuate or move underground. In Vinh City, bombers pummelled infrastructure, requiring the dispersal of industry to remote areas. The state pharmaceutical company for the province of Nghệ An, named Xí nghiệp Dược phẩm,[8] followed suit, relocating its production and storage facilities to the mountainous district of Đô Lương in the commune of Giang Sơn. Travelling by night, work brigades transferred over 1,000 tonnes of machinery and supplies to the secret location in an effort to continue the manufacture of much-needed medications for the ill and war-wounded (Trần 1989, 4). Unlike other enterprises that returned to the city after the air war ended in January 1973, CT Dược's central production operations remained in Đô Lương until 1981, with its workforce obliged to stay in collective, single-storey – or level 4 (nhà cấp 4) – housing in cramped and rudimentary conditions.

Despite critical disruptions to manufacturing during the period of evacuation, officials reported that workers and technicians at CT Dược continued to exceed production quotas between 1965 and 1979.[9] Like Tine Gammeltoft's (2014, 105) observation of health care workers in Hanoi, in the local press pharmaceutical employees were portrayed as "heroic individuals who, even during armed conflict, worked selflessly for the people's health." Charged with the task of producing medicine to keep the nation strong, the few hundred employees laboured industriously in difficult, mountainous terrain to overcome wartime shortages through the discovery and cultivation of new medicinal plants.[10] They also fought as militia forces to thwart approaching attacks. In the eyes of the state, the pharmaceutical industry had made exemplary contributions to the war effort in both production and defence of Military Area 4.[11] In recognition of these achievements in the face of great adversity, the president of Vietnam conferred the status of "labour medal, class 3" (huân chương lao động hạng ba) on the company in 1981 (P.V. 1981, 4).

It was against this backdrop of pharmaceutical heroism that housing allocation at CT Dược took place. Scholars such as Nancy Kwak (2015) have argued that the post-Second World War era saw housing emerge as a strategic Cold War instrument of policy and power. Housing estates, in particular, embodied the quintessential "rationalist vision" of utopian modernism across Latin America, the United States, and Europe (McGuirk 2014, 9). In the Soviet bloc, the right to affordable housing was pivotal to the socialist vision of prosperity and equality (Harris 2013). Similarly, in Vietnam after the First Indochina War (1945–54) overthrew French rule, state authorities had their own housing questions and crisis to resolve. Although the right to housing was enshrined in the Vietnamese Constitution, there was no centralized governmental response in Vietnam to the critical need to house a growing population, unlike the mass construction of prefabricated Khrushchyovka flats in the Soviet Union in the 1960s or the Wohnungsbauprogramm (Mass Housing

Program) in East Germany in the 1970s. Rather, postcolonial housing policy was oriented towards the Chinese model based on the *danwei*, or place of employment, which was responsible for the basic needs of workers and their families (Bray 2005). This reflected the organization of socialist society into labour collectives or work brigades in state enterprises and cooperatives, which were allocated either seized housing stock (in places like Hanoi) or tracts of land by the local government with the aim to increase productivity and self-sufficiency given critical shortages of food and material resources at the time. As in other socialist countries, the Vietnamese leadership prescribed to a belief in the abolition of capitalism and the redistribution of land and property as a solution to the housing question (Engels 1979 [1872]).

During and immediately after the war, work brigades shouldered responsibility for housing construction. In interviews, retired state employees in Vinh City – including former university faculty, cadres at the Department of Culture, or workers at the provincial confectionary factory – recollected how their brigades had collectively built simple *nhà tập thể*, or communal housing with shared facilities, made of wood, thatch, and bamboo. Retired workers from CT Dược, in particular, recalled the overcrowded living conditions in Đô Lương in wartime (single rooms were commonly occupied by several individuals or families) and their exposure to the elements: winters were cold, with the winds and rain penetrating their rudimentary structures. Not all workers lived in the nhà tập thể during the evacuation, however. Single employees, as well as families with absent husbands who had gone to war were usually assigned to collective housing, the vast majority of whom were low-skilled female workers. Other families, with male heads of households – typically managers – lived in provisional, self-built structures, or *nhà tạm*, made of wood and covered with thatch. According to one man named Mr Thành, nhà tạm were randomly built and scattered around the collective facilities, adjacent to the workplace. "At that time, people could claim any piece of land that belonged to the commons," he explained, but only if they had the means to build their own shelter.[12] In hindsight, it is possible to see how this seemingly practical arrangement, based on social divisions of labour in the workplace (managers were not sent to war, unlike the husbands of low-skilled workers), laid the foundation for gender disparities in access to housing insofar as a claim to land in wartime implicitly required the presence of male labour in the household. As Thành recollected, male family members, including husbands and older sons, would venture into the forests to cut down trees and gather bamboo to make rafters that could support the thatch roof and walls. "Women couldn't do this difficult labour alone," he claimed, drawing on essentialist notions of biological difference between men and women. Later, these nhà tạm would serve as a form of valued security – as proof of a model family's self-sufficiency – when faced with the prospect of returning to Vinh.

Return to the City: Social Housing and the Promise of Modernity

Differential housing allocation during wartime based on determinants such as need (associated with women and their desire for protection) and capability (associated with men and their ability to care for themselves and family) led to gender disparities in the distribution of postwar housing with relocation back to the city in 1981. Like other state enterprises faced with a growing workforce, the pharmaceutical company received a tract of land for its operations, with additional parcels for distribution among qualifying employees. It also received block C8 in Quang Trung, which was completed in 1980 at the end of the GDR's seven-year reconstruction project. Despite the attempt to meet the housing needs of its employees, shortages remained rampant, which led to unequal distribution among the meritorious, as also noted of housing policy in postwar eastern Europe.[13] Allocation of housing in socialist societies was typically based on need and merit, the latter of which frequently took precedence over the former. In Vinh City, workers were assigned a numerical value according to activities performed in the revolution (hoạt động cách mạng), excellence in labour performance (lao động xuất sắc), size of family, and housing urgency, among other criteria, with higher values resulting in "priority" (ưu tiên) status. At CT Dược, the point system not only determined who would receive one of three options – collective housing (the least popular option), land, or an apartment – but also allocation within each category (parcel size or apartment floor, for instance, lower being more desirable). Through these divisions, which mapped onto gender hierarchies, social stratification became more discernible in the built environment over the long term.

In interviews, most workers – the majority of whom were rural migrants – expressed their desire for a parcel of land over an apartment, which they considered unsuitable (không hợp) for Vietnamese lifestyles: the blocks were too high (people wanted to live on the ground floor, which was reserved for higher-level cadres), too loud, and too crowded. There was no individual courtyard for private gardens (these came later).[14] But land typically went to those with higher merit, such as men in management positions, rather than those with need – namely, female lower-skilled workers with children. This policy outcome reflected

gendered practices of housing allocation already in place that privileged male-headed households over female-headed ones, since to
qualify for land employees had to demonstrate they had the material
resources to build their own dwelling. In other words, how one lived in
the mountains during evacuation shaped to a large degree housing distribution with return to the city. As Thành explained: "The *nhà tạm* provided a kind of assurance that families would not live without shelter,"
since housing was not only a right but also an obligation. Given the
critical shortage of construction materials at the time, cadres like Thành
dismantled their makeshift homes and brought the wooden frames
with them to Vinh City to settle in the outlying district of Hà Huy Tập,
inscribing a pattern of gender segregation on the postwar cityscape.

Ms Thủy, on the other hand, had been allocated an apartment in block
C8 in the city centre. In contrast to Thành, she began her career as a
worker (*công nhân*) at CT Dược in 1968 at the height of the air war.[15] She
was the first in her family of rice farmers to leave agriculture for a job in
a state factory, which eventually led her to settle in the city. "It was easy
to get a job in those days," she laughed. ("Unlike today!" her husband
chimed in). Though Thủy faced strict discipline in the workplace – for
example, she was instructed to show up punctually every day – she
remembered the work environment as imbued with deep sentimental-
ity (*tình tâm lắm*) given the number of female youth employed by the
company at the time. When I asked her about the "land or house" decision, she thought for a moment and answered:

> I was given an apartment for me and my three children. My husband
> was away in the military at the time and I didn't have the opportunity
> to build a home. So I didn't qualify for land ... But I liked the thought of
> moving into a modern apartment. In Giang Sơn, communal housing had
> been crowded. It was made of thatch, and flooded in the rainy season. The
> apartment gave me a higher standard of living (*mức sống cao hơn*): it was
> more hygienic (*vệ sinh*) and civilized (*văn minh*). It was a new life for me
> (*cuộc sống mới*) – the first time I lived in a brick building with a private
> lavatory!

As Thủy suggested, in their modernist design and architectural
form the East German-built housing blocks seemed to offer residents a
path to urban modernity that embodied the hope of postwar recovery,
a theme that underpinned narrative reflections on that time. Another
female resident similarly commented: "Life in collective housing during evacuation was gruelling (*khổ*). We were better off in the high-rises
(*nhà cao tầng*)." Officials encouraged such sentiments, along with the

press, which praised the self-contained apartments as desirable, convenient, and ideal for workers and their families. Household amenities, like the indoor plumbing that excited Thủy, were to replace the drudgery of collective living; the reproduction of labour power would instead take place in the privacy of the single-family home.[16] And yet this enchantment with socialist modernity quickly dispelled after tenants moved in and infrastructure, such as water and electricity, did not function as planned (Schwenkel 2015). Block C8 faced yet another critical issue: rather than assign one family per unit, as per the master design, Vietnamese authorities proposed *two* families per flat, much to the dismay of GDR planners and residents alike.

In the original plan, the layout of the apartments was fairly standardized across the complex: units ranged between thirty and fifty square metres, depending on family size. Apartments had a consistent, utilitarian division of space to fit with infrastructural needs: each had a toilet, washroom, and adjacent kitchen on one side, and a living area and separate bedroom (in most units) on the other. The last blocks built (C7, C8, and C9) deviated from the conventional design and modernist ideal of one family per unit. During fieldwork, these three blocks were disdained as a disastrous adaptation of the standard built form. Changes advocated by the Department of Construction were largely instrumental, intended to save money by cutting costs (e.g., omitting balconies) and housing more people. The redesign created tensions between Vietnamese officials and GDR engineers. In contrast to the rationalized design of space in the other blocks, such modifications, GDR technicians warned, would result in unsanitary and overcrowded units that would accelerate the buildings' deterioration before their projected use life of eighty years. And yet, Vietnamese planners saw the redesign as more rational, if not ethical, given that allocation to more than one household helped with the acute need to provide shelter to an expanding number of employees. Authorities proceeded with their plan and after completion of construction allocated forty apartments in C8 to eighty female workers and their families.

When Thủy moved in, four people occupied her room (Thủy and her three children); in the other room lived a couple with one child, for a total of seven people in the thirty-square-metre flat. This was actually on the lower side of the occupancy rate, which grew over time to three generations. The women I interviewed did not imagine they would be forced to share the apartments for any length of time; each one thought it was a temporary fix until additional company housing was built. Over time, as some families moved out, the remaining tenants came to occupy the entire flat. In 1990, Thủy moved to another apartment after a workplace accident secured her a single-family unit. But hers was a special case. When I arrived to conduct fieldwork in 2010, eight of the units remained occupied by two families who had been living together for more than thirty years.

Divergent Paths: Revisiting the Housing Question after Đổi mới

The decision to double the number of occupants in C8 continues to anger residents who see the government's unilateral action as having produced hazardous living conditions. Others have blamed the "uncultured" (*vô văn hoá*) women themselves for the rapid deterioration of social housing, revealing longstanding tensions in social housing between rural migrants and those who imagine themselves to have more urban sensibilities (Schwenkel, forthcoming). Regardless of the reasons for the building's structural issues, it is widely recognized that C8 tenants live in the poorest and most precarious housing in Quang Trung, as opposed to the managers with land who have built stable homes and profited from a flourishing property market. In the next sections, I show how the lives of female workers in apartments and male managers with land have diverged with economic reforms and post-reform housing policy that promoted the privatization of state property. While the former is an account of decay and stagnant poverty, the latter is a narrative of growth and relative prosperity.

Housing: Life in Quang Trung

In Quang Trung housing estate, "decay" (*xuống cấp*, literally: to decline in grade or status) emerged as a metaphorical condition of material vulnerability that marked the women's embodied experiences of urban poverty. During my first visit to block C8 in November 2010, I received word that some residents wanted to meet with me to discuss their living situation. As I climbed the dank and decrepit stairway, a group of women gathered around me, pointing to the severe deterioration of the building. A bamboo pole had been positioned to keep a cement slab from the front corridor from dropping onto the courtyard three storeys below. The concrete pillars had completely eroded, some were haphazardly held together by thin wire, and large plaster pieces of the ceiling had fallen, exposing rusted rebar and prompting residents to build a makeshift cover to catch falling debris (see figure 1.2). Such do-it-yourself repairs have long been a part of collective life in Quang Trung, given the state's neglect of regular maintenance of the housing

1.2 Crumbling hallway and makeshift cover to protect residents in block C8, Quang Trung, 2010. Photo by Christina Schwenkel.

blocks (Schwenkel 2015). I had listened to countless stories about chronic disrepair during my research, and even read about it in the local press – for example, how a woman in C8 had been knocked unconscious by a large piece of dislodged cement, forcing the residents to wear conical hats in the apartments for protection. During interviews, residents explained how hazardous conditions in the home – from water leaks to cracked walls and ceilings – adversely affected their everyday domestic activities as well as their health and wellbeing. They consistently expressed fear that the building would collapse (sợ sẽ sập), a sentiment bolstered by a report released by the Department of Construction stating that less than 40 per cent of the building's structure remained intact. Indeed, these precarious living conditions were acknowledged by the state: soon after I began fieldwork, Typhoon Megi prompted city officials to order evacuation out of concern that the building would buckle from high winds.

Anxieties about safety and risk management were also on the minds of two women I visited, who had been sharing an apartment for over three decades. From the corridor, the house number was the only visual sign of the division of the unit: 131A/B.[17] Walking in, I observed the washroom across from the kitchen, as per the Vietnamese reform of the master plan. The family in room A to the right had laid claim to the kitchen, leaving the family in room B to the washroom, which served as their cooking area. Only the toilet was shared space. And yet the families moved fluidly across the space during my visits as if they were members of the same kin group. The women, both of whom had user rights in their name, invited me to sit down on a reed mat. This was the first time they had hosted a foreigner, and they were embarrassed at not having better amenities. The sparseness of the flat and its degradation (nothing had been renovated, unlike units in other blocks I regularly visited) indicated the relative poverty of its inhabitants. In the room where we sat, the only pieces of furniture were a wooden chest and a bunk bed to provide more space to accommodate three generations. I noticed irregular patches on the ceiling, next to widening cracks, as if the tenants had tried to close the gaps and gave up. "The apartment is not safe," the woman from room B said, following my gaze. She, too, had been injured by loose plaster that broke off the ceiling, but housing authorities did nothing to repair the damage. She broke down: "How can I live in a home that is danger- ous? This apartment has degraded beyond repair!" The local press took action, however, and published an article about the incident, exposing the state's indifference in an effort to force its hand (Mỹ Hà 2011).

Both women were also anxious about the proposed demolition and reconstruction of C8 as part of the city's plan for urban renewal, as well as its decision to privatize state property as per Decree 61/ CP. Although it would give the women shared ownership rights to the apartment, and the right to negotiate fair compensation from the developer, neither could raise sufficient funds for the legal transfer of the unit. "We are now forced to buy this run-down apartment from the state, even though we have paid rent for thirty years!" one woman complained. This was a heated topic across the complex. In C8 131A/B, each retiree paid approximately 15,000 Vietnam đồng (VNĐ; approx. US$0.75) per month for rent. Privatization would cost each household 10 million VNĐ (US$500) – an exorbitant amount, I was told, given monthly pensions of US$80. This created a dilemma for the women: if they did not agree to purchase their decayed, shared unit from the state (which they felt they already owned), they would not have a place to live; if they did acquiesce (as 98 per cent of tenants eventually did), they would then be obligated to buy into the newly planned condominiums at a rate of compensation agreed to by all sides. Both women were also exasperated by the latest news: the developer had proposed relocat- ing families in C8 to temporary housing in the newest twenty-storey

1.3 Handico tower, proposed site for the relocation of C8 families, butting up against block C6 in Quang Trung, 2011. Photo by Christina Schwenkel.

high-rise adjacent to their building – with two families per apartment (see figure 1.3).

This rumour also enraged the downstairs neighbour, Ms Ngọc, whose story deviated from the others owing to her more privileged background. A native of Vinh City, Ngọc received her degree from the Pharmaceutical University in Hanoi in 1974 and returned to Nghệ An to work at CT Dược. Given her credentials, she was assigned to the technical department, which gave her a higher salary than the lower-skilled workers who were her neighbours (only three technicians were in C8, she informed me). Ngọc came from a prominent family; her father had once been a high-ranking official at the Provincial People's Committee. In 1980, when faced with the decision of land or housing, she and her husband, an employee at the state electrical company, took housing out of concern that they would not have the means to build a shelter for their three children. The couple also subscribed enthusiastically to the image of socialist modernity, which for them was exemplified by the GDR-planned high-rise buildings. Ngọc was excited to move out of collective housing in the mountains and into a modern apartment in the city centre. She longed for a stable home where she would not have to worry about storms (though ironically, she had just been evacuated during the typhoon). For Ngọc, the housing estate offered modern conveniences, though it was crowded and the infrastructure was unstable. When she moved in, eleven people shared the split unit: six in room A (a mother and five children; the father had been killed in war), while Ngọc's nuclear family had five members. Like in 131A/B, the families coexisted by dividing the communal facilities: Ngọc's family had the kitchen, while the other family turned the washroom into their cooking area. In 1992, the tenants in room A moved out and Ngọc's family, now extended to three generations, has since been able to keep the unit to themselves. Like the women in the shared apartment upstairs, the topic of privatization and redevelopment infuriated Ngọc. She too complained that despite her regular payment of rent over the past years (36,000 VNĐ, less than US$2), she had to come up with 17 million VNĐ to assume legal ownership, close to US$850.[18] Ngọc also worried that relocation would force the dispersal of their tight-knit "family" of retirees from CT Dược: "It would be nice if we could stay together; that's how we have lived for several decades now. We know and understand one another," she said, emphasizing the modernist housing block as an affective community.[19] Ngọc was likewise dismayed by the rumour that the developer planned to resettle two families per unit in the neighbouring residential tower, on the highest floors no less, which was less desirable and had a lower property value (though this is starting to change with luxury dwellings). "Two families – what a scam (*bị lừa*)! After thirty years of sharing!" a neighbour angrily proclaimed from the doorway as she listened in on our conversation.

Land: Life in Hà Huy Tập

The post-reform experience with housing for the male managers granted land stood in sharp contrast to the women's experience with "xuống cấp" in the apartments, with material as well as economic and even social decline. I arrived in the district of Hà Huy Tập, where CT Dược had allocated land to its higher ranked employees, on a warm spring day in April 2011. The lane off the main road that leads to the settlement was narrow and bordered by stone fences that enclosed small parcels of property. Just beyond the tall buildings that line the busy thoroughfare of Lenin Road, Hà Huy Tập felt like a different world – quaint and peaceful like a small rural village. Chickens roamed among the modest, single-storey homes with courtyards and vegetable gardens – an image of urban life to which residents in C8 aspired.

In 1980, fifty-eight families were granted plots of land in Hà Huy Tập, which was then considered outside the city, three kilometres from Quang Trung housing in the centre. From the eastern edge, across a lush field of water ferns, there is a clear view of a new affluent community sprouting up, with large, multi-storeyed homes. While the economic disparity between these neighbourhoods is unmistakable, the CT Dược retirees in Hà Huy Tập have still fared much better than their colleagues in Quang Trung. Rather than progressive decline and disrepair, here one find signs of steady growth and even prosperity. But it took many years for such change to occur.

The homeowners of Hà Huy Tập I spoke with had been less enchanted with the promise of modernity bestowed through socialist housing: all had preferred land over an apartment. They were leery of living in a high-rise (especially the need to climb stairs), and desired a simple, stand-alone house with outdoor space for gardening. The obligation to share a unit was also off-putting. And yet, residents in Hà Huy Tập faced grave difficulties after relocation back to the city, and several people wondered if life would have been easier in Quang Trung housing at the time. Mr Bình, for example, now eighty-six old, lives on a 180-square-metre plot that he received in 1980. He had worked in management, like his fellow male cadres who received land. For five years, Bình and his family lived in a *nhà tranh* (thatch house) with a timber frame until they were able to build a more permanent home of sand-lime blocks. They also had to raise the foundation by one metre after the mass flooding that occurred in 1982 ("Water up to my chest," one woman remembered). These difficult early years gave way to new opportunities to improve living conditions and extend living space after Đổi mới reforms: In 1991, as construction materials became available for purchase on the market, Bình destroyed the limestone structure and built a more solid (*kiên cố*) house out of steel, cement, and brick, which still stands today. Since then he has expanded the house from two to four rooms – from 60 to 100 square metres.

Land was not equally distributed across management, however. Like in Quang Trung, size and location reflected rank and hierarchy in the company, as well as one's priority status. Because Bình's neighbour, Mr Trâm, held a higher-ranked position at CT Dược, his plot of land was larger (200 square metres) and situated in a prime location, closer to the road on the edge of the settlement. When he came over to introduce himself, he told me bluntly, "Women were in worker positions and were allocated housing, while we [male superiors] were given land," pointing to the relationship between the gendered division of labour and disparities in socialist housing policy at CT Dược.[20]

1.4 A single-storey home in Hà Huy Tập, Vinh City, April 2011. Photo by Christina Schwenkel.

Like Bình (and unlike the women in C8), Trâm's housing situation had similarly improved after economic reforms and the growth of the commodity market. Trâm proudly showed me his newly refurbished home, which also measured 100 square metres, with two bedrooms, a central guest area (with air conditioning), kitchen, and a long hallway separating dry and wet areas that led to a washroom and lavatory. Behind the house stood a row of verdant banana trees, while the front was adjoined by a quaint courtyard with newly laid tiles and an abundant vegetable garden (see figure 1.4). "It's a play area for my grandchildren," Trâm said as he recounted the material history of his house to describe a personal journey of overcoming adversity. Like others in the neighbourhood, in 1980 Trâm and his family built a simple thatch dwelling, with a cooking area out front and a latrine in the back.[21] The first years were full of hardships, he recalled. Given their distance from the city centre, there were no public services, such as water and electricity. They dug wells, and hauled earth to prevent flooding. "In the early years it was probably easier to live in Quang Trung," he surmised, though residents there faced similar difficulties with the breakdown of infrastructure. This gradually changed. By the time they started to build more permanent structures, most people had electric power in their homes. Today,

Trâm, Bình, and their neighbours live a modest and more comfortable life than their colleagues in social housing. They are not burdened with anxiety about safety, dispersal of their community, or demolition of their homes. Though not wealthy by any means, their household stability, compared to the precarious lives of female tenants in state-managed apartments, suggests that over the long term the higher ranked, male employees who received land were better off looking after their own housing interests. They also stood to benefit more from privatization.

Residents in Hà Huy Tập were not only able to stabilize their housing situation after construction materials were abundant on the market. With the Land Law of 1993 that permitted the legal transfer (or sale) of user rights, many were also able to achieve a degree of material prosperity. Land that once had no economic value (at least, officially), suddenly became desired and costly private property. An emerging real estate market that placed a premium on private home ownership created profits for many, including in Hà Huy Tập.[22] Mr Trâm's neighbour, for example, sold one-half of his plot (eighty square metres) at $3 million VNĐ a square metre (US$150) and invested his returns (240 million VNĐ, or US$12,000) into building a two-storey house. When I expressed surprise at the price, Trâm responded: "That's cheap! Land in this area of the city is inexpensive because there are no access roads for cars, just paths for motorbikes."[23] Another man who joined us lived on the periphery of Hà Huy Tập, where the street widened and became accessible by larger vehicles. Plots in his area tended to be larger – between 200 and 230 square metres, suggesting that he held an even higher position at CT Dược. Land prices in his corner were more exorbitant and went as high as 9 million VNĐ, or US$450, per square metre – just under the amount that each woman in 131A/B was struggling to raise to assume joint ownership of their crumbling, thirty-six square metre apartment. This man's comparison of the land situation in 1980 with the current pace of urban growth could not have been more different from the experience of the women in C8: "At that time, land had no value, unlike today when some of us have sold part of our plot to build a larger house. You can see for yourself that Hà Huy Tập now has some of the nicest homes in the city!"

Conclusion: The Demise of Social Housing

It has long been recognized that privatization of social housing has increased urban poverty and strengthened longstanding inequalities in market socialist and post-socialist countries across social and ethnic groups.[24] Low-skilled workers without social and cultural capital have tended to fare the worst. Tran and Dalholm (2005) have shown this to be the case in Hanoi, where post-reform policies have worked to the benefit of senior officials while neglecting to bring significant improvements to low-income households. A comparison of the lives of CT Dược retirees in block C8 (workers) with those allocated land in Hà Huy Tập (managers) in Vinh City confirms their observations. Absent from this discussion, however, is recognition that low-skilled workers under socialism, under the guise of emancipation, were disproportionately women (Gal and Kligman 2000, 8). It is therefore imperative to bring gender to the centre of the housing question, especially to the issue of urban housing reform in Vietnam (Hoang 1999, 77).

At the time of my research, more than one hundred nhà tập thể for workers continued to house urban residents in Vinh City (Hoàng 2011). Most of these collective accommodations were built during the subsidy years of state socialism in the "nhà cấp 4" or single-storey style. The Quang Trung estate, on the other hand, remains the oldest and largest social housing settlement in the city – and the only one built with foreign assistance, which rendered it modern and future-oriented at the time. According to the Department of Construction, all social housing in Vinh City is in a state of serious disrepair (xuống cấp trầm trọng). And while the post-reform housing question has led to a number of innovative solutions, including community-driven planning, in most cases the problem of housing has been left to the market. As seen in the examples of (male) cadres allocated land and (female) workers allocated housing, reliance on the private sector has deepened inequalities built on pre-existing gender hierarchies and biological essentialisms that were exacerbated rather than eradicated under socialism. Paternalistic housing policies meant to protect women considered vulnerable without male heads of household, for example, were shaped by ideas about "natural" differences between men and women. These same policies, in the long run, profited male cadres, who were considered by the state to be stronger and more capable of settling a new urban frontier.

Nonetheless, privatization of social housing has turned the women in C8 into legal homeowners, even as their property has been slated for demolition. This form of lawful ownership, in contrast to their sense of rightful possession through decades of occupancy and rent payment, has allowed them to strengthen their demands and play a more pivotal role in urban redevelopment, given that all plans require the approval of a two-thirds majority of tenants. In February 2011, C8 residents successfully negotiated with developers to increase the rate of in-kind compensation from 1.3 to 1.5 square metres.[25] Later, after rumours that resettlement would involve the temporary placement of two families in one apartment, they raised their demands. Like their landed colleagues, the retired workers in C8 have also recognized the market value of their collective property. Regardless of the building's material decay, it stands at a major intersection on land that commands a high price, which developers have been keen to acquire. In a petition filed with the province in August 2011, residents rejected the offer of higher compensation citing the favourable location of C8 on a main road in the city centre that would allow the investor to profit from the sale of commercial real estate. As an unexpected consequence of privatization, collective ownership of their shared property has placed the women in a more empowering position to influence policy and the terms of urban redevelopment (Schwenkel forthcoming).

The press has continued to follow the story of C8, framing the appalling situation of the elderly women as labour heroes who devoted their lives to building and defending the nation, who were showcased as socialist moderns living comfortably in high-rise housing and are now the face of the new urban poor. This portrayal of celebrated proletarians turned pitied precariat has made developers uneasy, and they have been careful to craft their public image of working compassionately with the resolute tenants, pledging to resolve their unfortunate plight in the best of their interests. But how the "best interests" of the residents will intersect with the profit-oriented goals of the developers and the state has yet to be determined. As of my last visit in December 2016, the residents were still living in the decrepit housing block. The renewal project stalled with the global financial crisis and new investors have been leery of the women's refusal to renegotiate their demands. For them, the right to housing, as promised by the state and enshrined in the constitution, is non-negotiable. At the same time, some neighbours in Quang Trung have grown critical of the women, even resorting to anti-rural bias by accusing them of being foolish about their entitlements or outright naïve about the housing market, even though the valuation of their property shows careful economic forecasting. In the end, these types of inflammatory gendered rhetorics effectively shift accountability from the state and its flawed housing policy to the female retirees themselves who, in their legal occupancy of the crumbling building, are reproached for holding up the renewal of the city.

NOTES

1 For more details on the housing complex and the history of its construction, see Schwenkel 2013. This chapter is based on one year of fieldwork in Vinh City in 2010–11, with follow up research conducted in Vinh City, Hanoi, and Berlin in 2012, 2014, 2015, and 2016.
2 See, especially, Wing (2005) on Shanghai, China; Pozniak (2013) on Nowa Huta, Poland; and Fehérváry (2013) on Dunaújváros, Hungary.

3 Not unlike the *danwei* system in China, which influenced Vietnam's housing policy.

4 Most research on housing reform and inequality has been carried out in the major metropolises of Hanoi and Ho Chi Minh City. See, for example, Gough and Tran (2009) or Labbé (2014) on Hanoi, and Harms (2016) on Ho Chi Minh City.

5 For similar observations in China, see Ho (2015) and Zhang (2010).

6 Such as the project to "cleanse" and depopulate cities under the Khmer Rouge (see Kiernan 2008, 31–64).

7 Because ownership technically lies with the state, I refer to land users, rather than owners here. Note that state-owned apartments were in the process of being privatized as per Decree 61/CP during my research in 2010–11, and by 2012 had been completed.

8 Later to become the *Công ty Dược phẩm* (shifting the name from pharmaceutical "enterprise" to "company"). I refer to the state institution as "Công ty [CT] Dược," as it was often called in conversation. Xí nghiệp Dược phẩm was founded in 1960 in accordance with Decision 134/QĐ-UB.

9 DNA Pharma, "Công ty Cổ phần Dược-Vật tư Y tế Nghệ An: 50 Năm Xây dựng và Phát triển 1960–2010" [Nghệ An Pharmaceutical-Medical Supplies Company Ltd: Fifty Years of Construction and Development, 1960–2010]. http://dnapharma.com.vn/News_.aspx?lang=vn&cat=25&id=568

10 The incorporation of traditional medicine into biomedical practices was rationalized as a necessity of war. By 1981, before operations returned to the city, 88 per cent of medicines produced by CT Dược were made from local ingredients (P.V. 1981, 4).

11 As per the three responsibilities movement where women's tasks focused on fighting, family, and production (defence and productive forces, as well as the affective labour of family care).

12 Interview, Vinh City, 27 April 2011.

13 For example, see Szelenyi (1983) for a similar critique of housing policy in socialist Hungary

14 For the creation of gardens out of green spaces designed for recreation, see Schwenkel (2017).

15 The story of Thủy's life is based on a series of conversations from November 2010 through May 2011.

16 This was far from actual practice. On the blurred lines between public and private spheres in Vietnam, see Drummond (2000).

17 I have changed the unit number to maintain anonymity.

18 Note that Ngọc's apartment cost more because she lives on the second floor (the lower the floor, the higher the market value at the time).

Deductions are also assessed according to contribution made to the revolution and number of years in state employment.

19 The story of Ngọc's life is based on a series of conversations from November 2010 through May 2011 that I also recount in Schwenkel (forthcoming).

20 A few female workers *were* able to acquire land, but through other means – for example, one woman lived in C8 for four years before she moved to Hà Huy Tập. She had applied for land but did not receive it. In 1984, she passed her apartment on to someone else and arranged a payment of 16,000 VNĐ, a fairly large sum at the time, for a 180-square-metre parcel.

21 The stove and the toilet were simple holes in the ground at that time, with ash from the wood stove, or *bếp củi*, used to process the waste.

22 For similar transformation in urban China, see Zhang (2010); on Eastern Europe see Fehérváry (2013).

23 Interview, Vinh City, 27 April 2011.

24 In Sofia (Bulgaria), for example, informal Roma settlements that were once tolerated under socialism have been deemed illegal and demolished, leading to a large contingent of displaced or even homeless residents (Ivancheva 2015).

25 As per Decision 926/QĐ-UBND, for each square metre of property owned, residents would receive 1.5 square metres in a new flat.

REFERENCES

Bray, David. 2005. *Social Space and Governance in Urban China: The Danwei System from Origins to Urban Reform*. Stanford, CA: Stanford University Press.

Collier, Stephen J. 2011. *Post-Soviet Social: Neoliberalism, Social Modernity, Biopolitics*. Princeton, NJ: Princeton University Press.

Drummond, Lisa. 2000. "Street Scenes: Practices of Public and Private Space in Urban Vietnam." *Urban Studies* 37, no. 12: 2377–91. https://doi.org/10.1080/00420980020002850

Engels, Frederick. 1979 [1872]. *The Housing Question*. Moscow, Russia: Progress Publishers.

Fehérváry, Krisztina. 2013. *Politics in Color and Concrete: Socialist Materialities and the Middle Class in Hungary*. Bloomington: Indiana University Press.

Gal, Susan, and Gail Kligman. 2000. *The Politics of Gender after Socialism*. Princeton, NJ: Princeton University Press.

Gammeltoft, Tine M. 2014. *Haunting Images: A Cultural Account of Selective Reproduction in Vietnam*. Berkeley: University of California Press.

Gough, Katherine V., and Hoai Anh Tran. 2009. "Changing Housing Policy in Vietnam: Emerging Inequalities in a Residential Area of Hanoi." *Cities* 26, no. 4: 175–86. https://doi.org/10.1016/j.cities.2009.03.001

Graham, Stephen. 2006. "Urban Metabolism as Target: Contemporary War as Forced Demodernization." In *The Nature of Cities: Urban Political Ecology and the Politics of Urban Metabolism*, edited by Nik Heynen, Maria Kaika, and Erik Swyngedouw, 234–54. New York, NY: Routledge.

Harms, Erik. 2016. *Luxury and Rubble: Civility and Dispossession in the New Saigon*. Berkeley: University of California Press.

Harris, Steven. 2013. *Communism on Tomorrow Street: Mass Housing and Everyday Life after Stalin*. Baltimore, MD: The Johns Hopkins University Press.

Ho Cheuk-Yuet. 2015. *Neo-Socialist Property Rights: The Predicament of Housing Ownership in China*. New York, NY: Lexington Books.

Hoang Thi Lich. 1999. "Women's Access to Housing in Hanoi." In *Women's Rights to House and Land: China, Laos, Vietnam*, edited by Irene Tinker and Gale Summerfield, 77–93. Boulder, CO: Lynne Reinner.

Hoàng Vĩnh. 2011. "Nhiều vướng mắc trong giải phóng mặt bằng" [Problems with Land Clearance]. *Nghệ An News*, 4 September. Accessed 12 December 2011 from http://baonghean.vn/news_detail.asp?newsid=76761&CatID=75

Ivancheva, Mariya. 2015. "From Informal to Illegal: Transforming Roma Housing in (Post-)Socialist Sofia." *Intersections: East European Journal of Society and Politics* 1, no. 4: 38–54. https://doi.org/10.17356/ieejsp.v1i4.82

Kiernan, Ben. 2008. *The Pol Pot Regime: Race, Power, and Genocide in Cambodia under the Khmer Rouge, 1975–79*. 3rd ed. New Haven, CT: Yale University Press.

Kwak, Nancy. 2015. *A World of Homeowners: American Power and the Politics of Housing Aid*. Chicago, IL: University of Chicago Press.

Labbé, Danielle. 2014. *Land Politics and Livelihoods on the Margins of Hanoi, 1920–2010*. Vancouver: University of British Columbia Press.

McGuirk, Justin. 2014. *Radical Cities: Across Latin America in Search of a New Architecture*. New York, NY: Verso.

Mỹ Hà. 2011. "Di dời nhà C8 Quang Trung (TP Vinh): Cần hài hòa 3 lợi ích" [Relocation of Block C8 in Quang Trung, Vinh City: The Need to Balance Three Benefits]. *Nghệ An News*, 2 November, 2.

Pozniak, Kinga. 2013. "Reinventing a Model Socialist Steel Town in the Neoliberal Economy: The Case of Nowa Huta, Poland." *City and Society* 25, no. 1: 113–34. https://doi.org/10.1111/ciso.12009

P.V. [Reporter]. 1981. "Đón nhận huân chương lao động hạng ba" [Welcoming the Labour Medal Third Class]. *Nghệ An News*, 26 May, 4.

Schwenkel, Christina. 2013. "Post/Socialist Affect: Ruination and Reconstruction of the Nation in Urban Vietnam." *Cultural Anthropology* 28, no. 2: 252–77. https://doi.org/10.1111/cuan.12003

– 2015. "Spectacular Infrastructure and Its Breakdown in Socialist Vietnam." *American Ethnologist* 42, no. 3: 520–34. https://doi.org/10.1111/amet.12145

– 2017. "Eco-Socialism and Green City Making in Postwar Vietnam." In *Places of Nature in Ecologies of Urbanism*, edited by Anne M. Rademacher and K. Sivaramakrishnan, 45–66. Hong Kong: Hong Kong University Press.

– Forthcoming. *Building Socialism: The Afterlife of East German Architecture in Urban Vietnam.* Durham, NC: Duke University Press.

Szelenyi, Ivan. 1983. *Urban Inequalities under Socialism.* Oxford, UK: Oxford University Press.

Trần Đức Thái. 1989. "30 năm xây dựng và phát triển của ngành dược tỉnh ta" [30 Years of Building and Developing the Pharmaceutical Field in Our Province]. *Nghệ Tĩnh News,* 15 May, 4.

Tran, Hoai Anh, and Elisabeth Dalholm. 2005. "Favoured Owners, Neglected Tenants: Privatisation of State Owned Housing in Hanoi." *Housing Studies* 20, no. 6: 897–929. https://doi.org/10.1080/02673030500291066

Werner, Jayne. 2009. *Gender, Household and State in Post-Revolutionary Vietnam.* New York, NY: Routledge.

Wing Chung Ho. 2005. "Negotiating Subalternity in a Former Socialist 'Model Community' in Shanghai: From 'Model Proletarians' to 'Society People.'" *The Asia Pacific Journal of Anthropology* 6, no. 2: 159–79. https://doi.org/10.1080/14442210500168283

Zavisca, Jane. 2012. *Housing in the New Russia.* Ithaca, NY: Cornell University Press.

Zhang, Li. 2010. *In Search of Paradise: Middle-Class Living in a Chinese Metropolis.* Ithaca, NY: Cornell University Press.

2 From ABC to Post-Industrial Suburb – Living in a Vision

BO LARSSON (TRANSLATED FROM SWEDISH BY AIDAN ALLEN)

Vällingby is perhaps the most discussed of all the suburbs in Stockholm. Since the inauguration of its centre in 1954 it has been visited, analysed, and written about innumerable times. The suburb has also been the subject of a dedicated monograph based on the results of comprehensive research (Sax 1998). Vällingby has been an interesting subject for the Stockholm City Museum, too. Our archive holds great amounts of material collected from the area, including building inventories, photographs, and other documents. In 2010, I interviewed people from eleven households, spread across various forms of housing in the area. The interviews followed up a study from the 1950s, when the scholar Börje Hanssen visited the suburb. Interested in what the transition of Sweden from an agricultural to an industrialized country meant for ordinary people, he put forward very prosaic questions (Hanssen 1978): What they were doing in and outside their homes, what kind of communications did they use, and what were their relations with neighbours, friends, and relatives?

In 2010 my purpose was to compare the 1950s with today, but in this chapter I aim to portray the everyday life in Vällingby and link it back to the idea of the "ABC town," which the area is seen to represent. The letters stand for *Arbete, Bostad, Centrum* (Work, Housing, Centre). People would not only live in the area; they would work, shop, and take part in social and cultural services here too. This was the fundamental concept behind the creation of the suburb, and one that I wish to analyse, to see if it is still relevant to Vällingby today. The suburb was intended for certain social groups: those we might broadly refer to as working and middle class, most likely families, preferably with children. My question is, therefore, What kinds of people live in the area today, and what is the nature of their everyday lives?

Many Swedish museums have documented contemporary life over the past thirty-five or forty years. Museum staff have located and interviewed people from varying walks of life and have also taken part in various activities themselves to acquire personal experiences. This recorded information, together with any collected objects, has been added to museum collections and often been disseminated in publications or exhibitions (see Silvén and Gudmundsson 2006 for an evaluation of the method).

Swedish museums have thereby gathered material for their archives and collections that deals with, and comes close to, individual people. This chapter forms part of that tradition. But to answer the specific question of this study I need to first sketch out a short history of Vällingby and the ideas that lay behind its creation. These include Swedish Social Democracy's ideas about society, given that the party completely dominated the political agenda before, during, and after Vällingby's creation.

When the Social Democratic party (SAP) came to power in 1932, Swedish housing standards were among the lowest in Europe, at the same time as rents were the highest (in this section I have drawn on Sax 1998, 21–7). Calls for improvement, first heard around 1900, culminated in the book-length manifesto *Acceptera* (Asplund et al. 1931). Its authors, established architects and art critics (among them Sven Markelius, to whom we shall return below), expressed a clear developmental optimism: technology could be harnessed to develop rational building methods to mass-produce cheap, good-quality housing.

Another influential book was published in 1934: *Kris i befolkningsfrågan* ("Crisis in the Population Question") by Alva and Gunnar Myrdal. (The couple were members of the era's intellectual elite. Gunnar Myrdal was famous for, among other things, his 1944 book *An American Dilemma: The Negro Problem and Modern Democracy*.) One of the book's proposals was adopted the following year: the building of slightly larger homes, with reduced rents, for families with children. In Stockholm two public-housing companies (a type of organization owned or controlled by the local authority, managing its housing stock on a non-profit basis) built around 2,000 homes before the Second World War, in Traneberg and Hammarbyhöjden, among other places. Here families with children enjoyed rent reductions of between 30 and 70 per cent, depending on the size of the family.

These new areas, built according to contemporary ideals, consisted of elongated residential blocks three to four storeys high, situated in open spaces, and set apart from industry and traffic. Yet they were soon criticized, among other things for being dormitory towns: they lacked workplaces and sufficient amenities, and communications with the city centre were poor. In response to this criticism, and inspired by Lewis

Mumford's famous 1938 book *The Culture of Cities*, the concept of neighbourhood planning was introduced when new residential areas were designed. Here Sven Markelius was a leading protagonist. A neighbourhood unit consisted of a clearly defined residential area, with a certain standard of amenities, and facilities such as schools. A number of neighbourhood units formed the basis for a larger common centre. Embedded in the concept was a vision of well-being, a sense of home, and community, which was expected to promote democratic life.

The new suburbs were also criticized for socially stigmatizing their inhabitants. *Barnrikehus* (residential blocks for large families) were quickly labelled "poverty homes." A 1945 analysis of social aspects of housing therefore concluded that government home-building subsidies should be universal. This, it was hoped, would prevent individuals and families with low incomes from ending up in specific "social housing." "This approach then became a characteristic of Swedish housing policy, in contrast to many other West European countries where often only low-income households were subsidized" (Sax 1998, 26).[1]

The latter half of the 1940s saw large-scale building of rental housing and cooperative flats across Sweden. The housing shortage was a major problem. At the end of the Second World War there were 2.1 million flats to house a population of 6.4 million people. However, at the beginning of the twentieth century, the City of Stockholm had already bought large areas of land. One such area would later provide the location for Vällingby. As newly appointed director of urban planning, Sven Markelius played a leading role in the design of the new suburb. In the publication *Det framtida Stockholm* ("Future Stockholm"; Markelius 1946) he presented ideas about the future form of the city and thoughts about ABC towns. Newly built suburbs would no longer be dormitory towns. Instead they would be fully provided with amenities and workplaces. The concept of the neighbourhood unit provided the blueprint.

The centre of Vällingby was inaugurated in 1954 and created a great deal of interest, so much so that long afterwards "area hostesses," tasked with receiving curious and enquiring visitors, were still employed (Sax 1998, 5).[2] The layout of the suburb is carefully planned, with the centre as a hub around which the various forms of housing are arranged. The most central of these is a string of eleven-storey blocks, intended to mark the limits of the centre and perhaps give a semblance of urbanity. Next, but no further than 500 metres from the centre, are three- or four-storey blocks of flats. These are arranged in small groups, each around a central garden, thus creating individual and secluded environments. Finally, northeast of the centre is a limited area of terraced houses and small detached homes. The three zones are interconnected by walkways and cycle paths, separated from road traffic. When the area was built it was estimated that some 42,000 people would live in its central parts, and that roughly half those of working age would be employed here (Creagh 2011, 16–18).

When the centre of Vällingby was inaugurated, the SAP held an almost hegemonic position in Sweden. By 1954 the party was halfway through a forty-four-year period of unbroken power (1932–76), even if it sometimes worked in coalition or collaboration with other parties. The party and its ideological aims greatly influenced the emergence of areas such as Vällingby. (The ABC principle, for example, also informed the design of Farsta, a suburb in the south of Stockholm that was inaugurated in 1960, and Kista in the north of Stockholm. Kista, completed in 1977, was the last suburb built according to the principle (Sax 2013, 10–11).

The SAP, when it was founded in 1889, was a party with strong Marxist influences that stressed the common ownership of the means of production. Although its alignment changed over the years (its more radical demands were tempered), at the end of the Second World War – in other words the era preceding Vällingby – the party put forward its postwar program. This included radical demands such as full employment, the equal distribution of wealth, an improved standard of living, and increased levels of democracy in trade and industry. The party had abandoned its previous categorical demand for the nationalization of the means of production, replacing it instead with an emphasis on a mixed economy (which became fully adopted in the party program in 1960). Pursuing full employment would mean, as far as necessary, the nationalization of natural resources, industrial companies, banks, public transport, and communications. For various reasons – including a period of unexpected economic prosperity, stubborn resistance from industry, and political opposition, but also the death of Prime Minister Per Albin Hansson – this policy was toned down or modified. However, the policy of full employment was pursued, and many social and cultural reforms were carried out.[3]

Its ideological development thus far can be traced in the party's own official sources. But historian Yvonne Hirdman, in an often-reprinted book, has observed the era from a gender perspective (Hirdman [1989] 2010). Her discussion of *folkhemmet* ("the people's home") and social engineering accurately reflects the post-war intellectual climate and thus provide another backdrop to the building of Vällingby.

The concept of folkhemmet, first used by the Social Democrats in the 1930s, characterized society as one single, great community, in which solidarity and helpfulness predominated. At the same time it presented an image of a reasonable, and thereby socialist, society where economic and social justice prevailed. This metaphor also signalled a sea change in Swedish Social Democracy: the emphasis shifting from the work-capital pair relationship to work-home. Many reasons lay behind this change – not least the poor standard of housing in Sweden – but the factor Hirdman stresses is ideological, a reorientation away from Marxist rhetoric (such as the seizing of the means of production) to a reforming model, "a 'reasonable' Social Democratic ideology of a planned economy" (Hirdman [1989] 2010, 94).

Setting the wheels of industry in motion first required consumption to be stimulated, and it was in the home that goods and services, made available through work, were consumed. Consumers, though, needed to buy "correctly": this would be informed consumption or, as Hirdman puts it, socialization carried out by consumers. If everyone were to demand a good, fit-for-purpose home – made up of at least two rooms and a kitchen – then the basis for capitalistic home building would be eliminated.

Given that the household was the woman's domain, women's issues became welfare issues too. This is where social engineering enters the picture: the question of reordering women's lives became a strategic issue in remodelling capitalist society. Knowledge and information became important means in this struggle.

The "social engineers" who Hirdman believes set the tone for the era were young, intellectual, radical, and modern. They are exemplified, she believes, by Alva and Gunnar Myrdal and their circle of friends who were economists and architects (among them Sven Markelius). This alliance or symbiosis of individuals and disciplines brought together economics, engineering, science, and politics. Primarily this meant a reinforcement of politics. When Hirdman summarizes the components of social engineering, she uses words such as *science*, *rationality*, and *objectivity*, but she also claims there was a will to achieve something and cites a confidence in the benevolent state and the potency of technology.

It would take too long to explore how this social engineering manifested itself in full, but the activities of *Hemmens Forskningsinstitut* (the Domestic Science Research Institute) make for a good illustration. This body, to a large part founded during the Second World War by two women's organizations (Kaijser and Sax 2013, 23) and funded by both state and industry, aimed to rationalize housekeeping via research and information. The institute made careful, closely detailed studies of everything from kitchen fixtures and domestic appliances to housework and children's home environments.[4] Hirdman's reflections on the institute are somewhat ironic, yet still informative: "For some reason I was greatly moved by the thorough inventory of three Swedish households, of everything from bedsheets to pots and pans, and the timed dressing of a five-year-old – not to mention the recording of the number of paces between stove and sink or the measurement of oxygen consumed while mopping the floor, effected by means of a special device fastened over the housewife's respiratory organ" (Hirdman [1989] 2010, 260).[5]

It is tempting to see both the allegory of folkhemmet and social engineering as background forces involved when homes in Vällingby were allocated. The Social Democrats wielded their political strength to achieve other goals too, for example to support large families (for population growth) and to encourage women to participate in the labour market. "The very viability," writes the architect and academic Lucy Creagh (2011, 20), "of Vällingby as an example of the 'ABC principle' was engineered by giving those who could find work in the area preferential housing allocation." In total 92 per cent of the homes consisted of rental flats, mainly two- or three-room dwellings (plus kitchen). But the variation in housing types fostered a degree of social stratification: the rental flats became home to workers and mid-range professionals, whereas the terraced houses and detached homes were the domain of higher professionals and academics (Creagh 2011, 20; Sax 1998, 63).

But the 1950s were not only a Social Democratic golden age; they were also a period that scholars have called Sweden's high industrial era (Isacson 2002). This period, which coincided roughly with the Social Democratic era, lasted from approximately 1930 to 1980. Today Sweden is no longer in its high industrial era and the extended period of Social Democratic dominance has long since ended. The 2010s are instead characterized by post- or late modernism or, in the view of some scholars, the Third Industrial Revolution (Ilshammar 2008, 254; Isacson and Morell 2002, 13). Even the ABC principle, so characteristic of Vällingby, appears to be gone. Vällingby Centre, the heart of the suburb, now goes by the name of Vällingby City, and in 2011 the website of its owner, Svenska Bostäder, rather than emphasizing the ABC principle, was promoting the clothing shop Kfem, "The latest star in Stockholm's fashion constellation" (Larsson 2011, 90).[6]

So how is life in Vällingby in this new era? Whatever name we use, the new era ought to influence the make-up of the population and its way of life. Of the eleven interviews I conducted in 2010, seven took place in Vällingby's three- or four-storey blocks (which are still rental homes). Because of space limitations I concentrate on these seven interviews. This study does not attempt any statistical conclusions but aims instead to provide qualitative insights into a modern way of life.

Workers and mid-range professionals originally made up the majority of housing-block tenants. This pattern was repeated among the

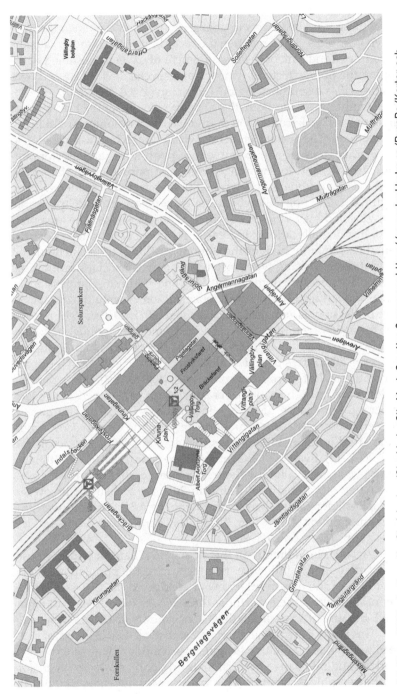

2.1 Map of Vällingby, from the Stadskartan (Stockholm City Map), Creative Commons: https://www.stockholm.se/ByggBo/Kartor-och -lantmateri/Kartproduktspecifikationer-/.

households I encountered a half century later. Of these, two are couples of old-age pensioners (formerly employed in industry and brewing in the case of the men, and in secretarial and kitchen work in the case of the women); two households consist of a single adult each (one a parent with several children), who are employed as a youth centre leader and a home-care worker respectively; in another household the woman combines her job as a personal assistant with degree-level economic studies, while her husband works in IT support. In the sixth household the man is a local government manager and the woman works as a receptionist at a large hospital, and in the seventh and final household the woman works in theatre lighting and the man is a product engineer for an international company.

The age range is wide: a difference of fifty-three years separates the oldest and youngest interviewee. The average age is fifty-two and, with the exception of one four-room flat, all the interviewees live in three-room flats. Given the relatively high average age, there were few children living at home, actually in only two of the households (although a mother in a third family was expecting her first child). The modern globalized world is reflected in household background – two families have roots in non-European countries: Eritrea and Iraq respectively.

All seven households are situated around the same enclosed garden. In the 1950s this area, and Vällingby as a whole, was considered new and fresh. The area hardly boasts this modern reputation any longer, but neither is today's visitor greeted by a half century of wear and tear. Instead, its buildings are sound and well maintained, and a recurrent theme in my interviews is how content people are with their living environment.

Two Families

In this section I describe, as far as space allows, two of the households – their backgrounds, life situations, and relationship to Vällingby. I hope this will provide a more in-depth image of the nature of present-day life in this very well-known Stockholm suburb.

Background

In 2010 Madelené lives on the second floor of one of the three-storey blocks that are typical of the second zone. She is about to turn thirty-six years old, and is a single mother with three daughters and a fourth child on the way. Her daughters, aged four, eight, and sixteen, are the main reason Madelené moved to Vällingby last year. She used to have a two-room flat in Husby, the northern suburb where she was born and raised, but her children, especially the teenager, needed more space. The choice of Vällingby stemmed from previous knowledge of the area: a niece lives here and Madelené had been to the area before, visiting some of the shops. This quiet area, home to many older people and families with children, is a good place to raise a family. In addition, a school and a daycare centre are located adjacent to her home.

August and Sara also live in a three-room flat, but on the first floor. Born the same year, they are both thirty-three years old and expecting their first child. They see their Vällingby flat as a practical step between the smallish flats of their youth and the little house in the Stockholm countryside they are eventually aiming for. Their previous home was Sara's little one-room flat in Högdalen in the south of Stockholm. From there August commuted weekly to the job he had at the time in another part of Sweden.

The couple moved to their Vällingby flat in 2009, the year before the interview. It was not hard for them to secure the contract: August had been in the housing queue for seven years. They thought this flat was the best one available, mainly because of the pleasant layout of the rooms and kitchen, all of which were of good size. This compensates for town being a thirty-minute metro journey away. They also wanted a relatively cheap flat – they pay just over SEK6000 (roughly CAD$840 in 2015) a month for their seventy square metres and garage space – and one that was situated to the south of the city. Otherwise August would find his journey to work took too long.

Work

The first time I arranged to interview Madelené I found a note on her door: "Hello dear interviewer. Really sorry but the planned interview is off! I was offered a shift and as a stand-in I can't say no. Need all the shifts I can get. My apologies." The note said much about her situation at the time. She was working as a personal assistant, a job she described as being someone's arms, legs, and sometimes voice too. This and similar home-care work has been her main occupation since she was eighteen years old.

She used to work varied hours, days and evenings as well as nights, which was difficult for her children and meant finding a babysitter. But when she moved to Vällingby she got the option to work nights. She thinks this is the perfect solution: she has time to deliver and collect her children from child care, as well as do the laundry and cleaning and everything else to do with having a family. Her work is within walking distance. It takes her between ten and fifteen minutes to get there.

However, the problem with the night shifts was that she did not sleep during the day and was sometimes awake for forty-eight hours at a time, not least because she wanted to spend time with her children when they were awake. This worked out, but she went around "like a zombie." Since then she has worked instead as a stand-in. This gives her some freedom of choice, but at times, when work is scarce, she must take what she can get. One such occasion was when she posted the note on the door for me. Later, when we finally got to meet, she was on sick leave with pregnancy difficulties.

Both Sara and August have links to the theatre, Sara through her job and August via his background. When interviewed, Sara was employed as a lighting engineer at a major theatre in Stockholm, but had for some time been on pregnancy leave. Otherwise she works short-term contracts, six months at a time. After having worked at a grocer's, among other things, at the age of twenty-five she started studying with a goal in mind: first a course in stage design, then another in theatre technology at the Dramatiska Institutet (DI). She graduated in 2005. She is not entirely happy with her choice of career. Her basic idea in attending DI was that she wanted to work as a lighting designer, but she now feels a little stuck in its technological aspects. However, she does not know if she has enough motivation to start studying again.

August works as a product engineer for a major international company. His previous employment included work as a theatre technician – which is something of a family tradition: both his parents were actors – and he has lived abroad to study languages. But municipally funded theatre work means working evenings, low pay, and tight budgets, and becomes wearing after a time. He therefore studied civil engineering at KTH Royal Institute of Technology in Stockholm. The program, including an extra year to complement his high-school studies, took five and a half years to complete. His work entails lots of travel: to Germany, Slovenia, and many other countries, but also – as at the time of the interview – daily commuting to closer destinations such as Uppsala.

Vällingby

Madelené describes Vällingby as a quiet place, at least compared to Husby, where she spent most of her life. In Husby things just got worse over time, and she saw how people she grew up with became drug addicts and petty criminals. She did not really expect Vällingby to be any better, but the situation here is quite the opposite. The place could not be quieter, at least in her area.

Vällingby has a wide range of different shops. Perhaps, though, she would like a shop like Willy's, a supermarket that stocks a range of quality, low-cost goods. She loves strolling around and looking in the shops in the centre, although she has been in only half of them. There is so much to look at and compare. In addition there is a library, swimming pool, and cinema. Her eight-year-old daughter likes reading, so they use the library. They have yet to visit the swimming pool or the cinema, and Madelené does not go to the cinema that often. But the range of choice in the centre is really nothing to complain about.

People in exclusive areas such as Danderyd and Djursholm almost certainly think Vällingby is a ghetto, quieter than Rinkeby and Tensta but, Madelené emphasizes, a ghetto nonetheless. Yet she thinks the area is good, much better than many other parts of town.

Sara and August like the area too, but would never have left the south side of the city had August not been offered a job there. The good thing about Vällingby is its variation, the green areas and different forms of housing – rented and privately owned flats, detached homes and terraced houses. The ABC principle resulted in a wide variety of inhabitants. Other parts of town where they lived before were much more homogeneous.

But Vällingby is Vällingby, not Stockholm. Travelling into town is often too much of a chore. They sometimes miss being closer, especially because most of their friends still live south of Stockholm. On the other hand, they never have to leave Vällingby unless they absolutely want to. The centre is good and as large as that in any small town, in terms of the number of shops at any rate. Vällingby feels better thought out, in terms of urban planning, than many other parts of town.

There are also many more green areas than elsewhere. They keep a small boat on Lake Mälaren and are, in their own words, quite good at filling a thermos and going out, just to sit by the water or on a park bench with hot chocolate and sandwiches.

Leisure

Madelené calls herself a typical animal person. Horses and dogs are her main interests. She has ridden since she was about five and has also leased a horse, which is effectively the same as ownership unless the animal falls ill. She pays for everything except health insurance. She has also competed in show jumping and pony harness racing. She has won a few competitions and beat some records. It was especially good

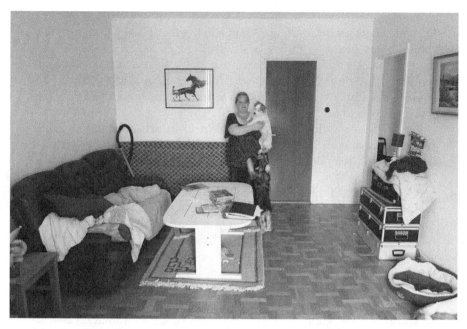

2.2 Madelené. Photo by the author.

when she beat the daughter of rock star Jerry Williams at show jumping. She does not like to boast but it was fun.

She still has a big interest in horses, something her daughters have
inherited, especially the oldest and youngest. On weekends they often
go with her to the horse she is currently tending. Her friend has a horse
at the same stables, an animal that is wonderfully gentle with everyone
and everything. The children can ride it. Her children go everywhere
she goes on weekends, when she tries to be with them "24/7." They like
sleeping with her in the living room.

Madelené has always had a dog and now she has two, one five years
and one six months old. The children love them, and Madelené could
not manage without them either. But they do demand commitment. The
older one manages with a walk three to four times a day, but the younger
one is being housetrained and needs to go out more often, sometimes at
night too. Occasionally they get a long walk of forty-five or fifty minutes,
but often they have to manage with shorter walks around the block or a
longer ramble in the immediate area. If Madelené takes the last walk of
the day late at night, the dogs will not need to go out until a little later
the next morning. She does not need to dash out with them.

In addition to animals, music – especially country music – and tattoos are Madelené's interests. When she lived in Husby, a country TV channel was included in the basic package, but then all the "brainboxes" decided to make it subscription-only. To watch it would have required a satellite dish, something Madelené could not afford. This made her really angry. It was her favourite channel and was always on when she was at home.

For Madelené tattoos are no different from art and they fascinate her. She has one on her arm: a heart, a rose, and the dates of birth of her three children: 1993, 2001, and 2005, each within its own pennant. When her fourth child is born a new date will be added.

If Madelené had the money she would tattoo herself more often. She would like to decorate both her arms and legs, even her collarbones – not her face, but the rest of her body. She really likes tribal and clan tattoos. But the process is not painless. Despite her high pain threshold, she suffered during each and every one of the hundred minutes her tattoos took to complete.

Over the past year, August and Sara have made time to go away for a few weekends. They have visited other towns, stayed at hotels or youth hostels, gone for a massage, and dined at upmarket restaurants. Sometimes they have also gone to a concert during a weekend away. The reason for these trips has been, in their own words, to take a break and feel good. They have also been abroad: Sara visited August when he was working in other countries – including Slovenia and Switzerland – and they have organized their own travels too. Last summer they embarked on a three-week car journey in Spain, which they were forced to cut short when Sara became ill with gall bladder problems. They had just learned she was pregnant, and did not want to take chances.

The couple owns a car, but August wants two. He would like a classic vehicle, to be driven simply for pleasure, and which unlike their everyday car would not always have to be on the road. At the moment they own a Ford Focus, an eco-friendly car that runs on ethanol. This means they avoid the congestion charge, but mainly they like the idea of supporting the eco-friendly car market. They used to drive a Volvo 745 GLE, a box with plush seats, which nostalgic August thought was wonderful. But it used too much gas for his conscience.

August also likes cycling, even though he probably did more of it a few years ago. A bike gets you everywhere and is a good cure for restlessness. A bike and backpack have always been his trademark, but now his job is too far away to reach by bicycle. He has considered buying a moped, which would avoid the congestion and still get him out

2.3 August and Sara. Photo by the author.

and about. He is looking for a vintage moped, preferably a Crescent Compact, which is both common and cheap.

Given Sara's profession, the couple often visits the theatre. Before her maternity leave they took advantage of free tickets to watch as many performances as they liked. The frequency of their visits depended on their working hours; sometimes they went three times a week. The couple likes movies too. Because there is no good DVD-rental shop in the area, they sometimes buy five or six discs at a time. August prefers more exclusive films – by Roy Andersson or Michael Haneke, for example – whereas Sara has broader tastes. At the cinema they prefer documentaries. Sara mentions *The Queen and I*, a film about Farah Diba, the former queen of Iran.

Jigsaw puzzles are another common interest. Sara is the driving force here, ensuring that the puzzles get finished, while August is more interested in the motif itself. They have to be kitschy, the more so the better, like deer or a Bavarian castle. Their idea is eventually to cover a whole wall with finished puzzles. Up til now they have done 1,000-piece puzzles; perhaps the next will be 2,000 pieces. It will cover a larger area.

The walls of the flat are also decorated with August's paintings. He used to paint much more, but has less time nowadays. Sara calls his motifs "odd little men," while August characterizes his style as figurative with a humorous feel. As far as he is concerned, it's OK for art to be pretentious so long as it has a twist.

Painting is not part of Sara's spare time. Instead she likes to read books, as many as possible. Her choice of subjects goes in phases, but for pure entertainment crime fiction is her genre. Otherwise her current focus is the history of religion. She became interested via *The God Delusion* by Richard Dawkins, a book that largely corresponds with her personal belief that religion only spreads misery. She then progressed to other books, wanting to learn about the "real" history of Christianity, not the version taught in school. She would love to write a dissertation on the books she has read as some form of project. She knows she will never get around to it, but the desire remains nevertheless.

The Future

Madelené enjoys her job, helping people. She always has; otherwise she would not be in this line of work. But when asked about the future she talks about her animals. Even if it might sound ridiculous to some, given that she is nearly thirty-six, she would like to work at an animal-rescue centre, and to live in the countryside with her children and her horses and dogs. She aims to reach her goal in ten years, perhaps even sooner if possible.

When I asked her to think of a symbol for her life, she chose neither her children nor animals. Instead she selected an axe, one of the ancient kinds associated with power. As a single mother she has done so much for her children, things that few people really think about. This affects the body and one ends up with economic problems. And there are many people like her who give their all, both for their children and to survive. This is why she does not choose a mobile phone or a computer or such; these do not represent strength to her. But an ancient axe does: it is an object associated with self-confidence and power.

When August and Sara talk about the future a slight good-natured disagreement breaks out. They agree about the house in the country, perhaps more children, and certainly cats. August will have his classic car, but when he says they will probably have the same or similar jobs, Sara reacts. What about everything else? What happened to Australia, which they used to talk about? She becomes a little disappointed with their outline for the future and doubts whether she would want to be doing the same job. But she agrees completely about the house. It would provide an outlet for her interest in gardening, which is quite new to her. At the moment she has a balcony, but it lacks the space for big projects.

As a symbol for her life Sara chooses a book, a detective novel. A book by a Swedish author who writes readable crime fiction – this is her and her life. August suggests his bicycle and a backpack. Working off restlessness through cycling has always been a big part of him.

Conclusion: The Varying Viewpoint

This chapter is based on interviews that were conducted with a somewhat different goal in mind (Larsson 2011). Because of this, in this chapter I have viewed them "obliquely." During the interviews I posed no direct questions about whether the ABC town still existed or not. Nevertheless, I still believe this question may be approached by using these conversations. A suitable starting point would be Lucy Creagh's (2011, 18) claims about variety itself as the leitmotif in the development of Vällingby. Although she refers primarily to the architecture and spatial experiences of the area's public domain, the same perspective may also be applied to social aspects over time. My description of the individuals who took part in the seven interviews suggests, for example, that variation over time is negligible in terms of who lives in the flats. The occupants are still the same types of people, who may broadly be called working and middle class.

In terms of household background and composition, diversity seems greater than it was in the 1950s. At that time, many families moved here from quite run-down parts of central Stockholm. Of the families interviewed in 2010 two had roots outside Europe (which is hardly surprising given the nature of global development). There are no longer really any nuclear families – mother, father, children – which Creagh describes as one of the Social Democratic goals in creating Vällingby. One family alone fits that description, with couples and single-person households being the norm. Average household age would also appear to be significantly older than when the area was new.

Present-day conditions fit better with the concept of people living and working in the same area: of the five working-age households, three are employed in and around Vällingby. But in contrast to the past, all the adults interviewed were working. The role of housewife – an ideal when Vällingby was built – no longer exists.

Thus this short summary of my relatively few interviews suggests the survival of some of the original intentions for Vällingby, despite the postmodern era. These are evident in the types of people who live in the flats, and in the fact that most interviewees work close to home. Can we see traces of the ABC principle in the close-up images presented here? I believe we can.

Madelené, Sara, and August all love the centre, which was built to gather all shopping choices in one location. Sara and August say that they never really have to leave Vällingby, because everything they need is there, and for Madelené the centre means more than consumption: she also enjoys just strolling around, window shopping. Madelené – who also works close to home – stresses that school and daycare are both situated nearby, and Sara and August actually refer directly to the ABC principle when they mention the area's diverse population. This is one reason they feel so at home here, which they attribute to the original variety of housing types.

Consumption and population diversity, and perhaps an appreciation of the suburb's green areas, have all been constants since the area was first built. The concept of variation need not be viewed only diachronically; it may also be combined with a synchronic perspective. Just as in the 1950s, working- and middle-class people live in the homes included in the study. Variation within these categories is sometimes wide. Madelené lives under fundamentally different conditions than Sara and August. Whether similar differences exist between other tenants in the area is hard to establish with certainty without a quantitive analysis of incomes and household size, yet such a state of affairs would not be unlikely given the evidence. The diverse make-up of the seven interviewed households would suggest so. Conversely, such major differences are unlikely to have existed when Vällingby was first built, when housing policy discouraged atypical households. Perhaps we should look at features of this kind to see real differences between then and now, and between present-day phenomena, too. This would demand a varying viewpoint, changing between long and short distance, between the detail and the whole.

NOTES

1 In Swedish: "Detta synsätt har sedan dess varit utmärkande för den svenska bostadspolitiken och skiljt den från många andra västeuropeiska länder som ofta bara subventionerat bostäder för hushåll med låga inkomster."
2 Miss Vällingby hostesses, as they were known, were employed by the main housing company and local entrepreneurs. Dressed in blue uniforms, the hostesses sold souvenirs and answered visitors' questions at a small information stand.

3 "Sveriges Socialdemokratiska arbetareparti," *Nationalencyklopedin.* https://
www.ne.se/uppslagsverk/encyklopedi/lång/sveriges-socialdemokratiska
-arbetareparti
4 "Hemmens Forskningsinstitut," *Nationalencyklopedin.* http://www.ne.se
/uppslagsverk/encyklopedi/lång/hemmens-forskningsinstitut
5 In Swedish: "Av någon anledning grep mig den stora inventeringen
av allt från lakan till kastruller hos tre svenska hushåll och tidtagnin-
gen av påklädning av en femåring särskilt hårt – och då hade jag ändå
inte studerat stegräkningen mellan spis och slask eller åtgången av
syre vid golvmoppning medelst en särskild apparat fäst över husmors
andningsorgan."
6 Recently, the name has changed another time: it is now Vällingby Centre
once again.

REFERENCES

Asplund, Gunnar, Wolter Gahn, Sven Markelius, Greger Paulsson, Eskil
Sundahl, and Uno Åhrén. 1931. *Acceptera.* Stockholm, Sweden: Tiden.
Creagh, Lucy. 2011. "From *Acceptera* to Vällingby: The Discourse on
Individuality and Community in Sweden (1931–54)." *FOOTPRINT – The
European Welfare State Project: Ideals, Politics, Cities and Buildings* 5, no. 2: 5–24.
Hanssen, Börje. 1978. *Familj, hushåll, släkt: En punktundersökning av miljö och
gruppaktivitet i en stockholmsk förort 1957 och 1972 enligt hypoteser som utfor-
mats efter kulturhistoriska studier.* Hedemora, Sweden: Gidlunds.
Hirdman, Yvonne. (1989) 2010. *Att lägga livet till rätta: Studier i svensk folkhems-
politik.* Stockholm, Sweden: Carlsson.
Ilshammar, Lars. 2008. "Det industriella tjänstesamhället." In *Industriland:
Tolv forskare om när Sverige blev modernt,* edited by Jan af Geijerstam, 247–71.
Stockholm, Sweden: Premiss.
Isacson, Mats. 2002. "Den högindustriella epoken." In *Industrialismens tid:
Ekonomisk-historiska perspektiv på svensk industriell omvandling under 200 år,*
edited by Mats Isacson and Mats Morell, 115–24. Stockholm, Sweden: SNS.
Isacson, Mats, and Mats Morell. 2002. *Industrialismens tid: Ekonomisk-historiska
perspektiv på svensk industriell omvandling under 200 år.* Stockholm, Sweden:
SNS.
Kaijser, Arne, and Ulrika Sax. 2013. *A Tribute to the Memory of Brita Åkerman
(1906–2006) and Carin Boalt (1912–1999).* Stockholm, Sweden: Royal Swedish
Academy of Engineering Sciences (IVA).
Larsson, Bo. 2011. "Vällingby 1957 och 2010: En återvändande undersökning."
In *Andra Stockholm: Liv, plats och identitet i storstaden,* edited by Bo Larsson
and Birgitta Svensson, 57–95. Stockholm, Sweden: Stockholmia förlag.

Markelius, Sven. 1946. *Det framtida Stockholm: Riktlinjer för Stockholms generalplan* [Stockholm in the Future: Principles of the Outline Plan of Stockholm]. Stockholm, Sweden: K.L. Beckmann.

Mumford, Lewis. 1938. *The Culture of Cities*. London, UK: Secker & Warburg.

Myrdal, Alva, and Gunnar Myrdal. 1934. *Kris i befolkningsfrågan* [Crisis in the Population Question]. Stockholm, Sweden: Albert Bonniers.

Myrdal, Gunnar. 1944. *An American Dilemma: the Negro Problem and Modern Democracy*. New York, USA: Harper & Brothers.

Sax, Ulrika. 1998. *Vällingby: Ett levande drama*. Stockholm, Sweden: Stockholmia förlag.

– 2013. *Kista: Den tudelade staden*. Stockholm, Sweden: Stockholmia förlag.

Silvén, Eva, and Gudmundsson, Magnus. 2006. *Samtiden som kulturarv: Svenska museers samtidsdokumentation 1975–2000* (översättningar till engelska: Alan Crozier). Stockholm, Sweden: Nordiska museets förlag.

3 The Rise and Fall of Collective Housing: Hanoi between Vision and Decision

LISA B.W. DRUMMOND AND NGUYỄN THANH BÌNH

As the capital of successive and ideologically opposed regimes – colonial, post-colonial, Communist, and "market socialist" – Hanoi has consistently grappled with a shortage of housing throughout its modern history. Colonial authorities answered the demand for housing with bouts of extending the city's land area by reclaiming swamps and ponds and swallowing up contiguous villages to create new districts for development. The Communist government of a newly independent Vietnam (the Democratic Republic of Vietnam) took a different approach, adopting the Soviet model of dense apartment blocks built in clusters (*mikroraion* in Russian, *khu tập thể* – KTT – in Vietnamese). The contemporary Communist-operating-as-"market socialist" government has turned to demolishing and then rebuilding those same KTT as condominiums. How far they have fallen! To be allocated a KTT apartment was, in the 1960s, to be offered the most modern and desirable living space in the capital. To live in a KTT apartment in 2018 is to be living under the shadow of the wrecking ball because the city authorities deride your home as an eyesore: a sign of urban backwardness, blight, and simply failure. What has happened to effect this stunning reversal of fortunes?

Urban Visions of the New Communist State and Their Consequences

As the capital city of the independent and Communist-run Democratic Republic of Vietnam, Hanoi grappled with a shortage of housing caused by an influx of cadre (loyal Party members) for the new Communist bureaucracy and workers for the new industries and factories. After the decisive Communist victory in 1954 and the retreat of the French, Hanoi's population experienced a short-lived reduction. French residents left, of course, but they had always been only a small fraction of the city's population. The bulk of the emigrants were Vietnamese Catholics or others who did not wish or feared to live in the Communist territory; they fled to the south, to the US-backed Republic of Vietnam and its capital, Saigon. This change represented only a minor dip in the capital's population, however, and within about five years the population had rebounded and almost tripled, from about 160,000 in the mid-1950s[1] to 458,090 in 1961 (Bùi Quang Tạo 1961)[2] in the inner city territory of twenty-eight square kilometres. According to Trịnh Hồng Triển (1984), then the deputy director of the Hanoi Construction Department, between 1954 and 1981 Hanoi's average per capita living space, including service areas, dropped to 3.27 square metres from 10, a rapid and dramatic decline. In 1954, Hanoi's total building stock was 2.27 million square metres, and in 1981 it was 6.95 million. Within those figures, housing stock accounted for 1.5 million square metres in 1954 and 3.45 million square metres in 1981. This represents a substantial increase over the three decades, but it had little impact on per capita living space due to even more substantial urban population increases, so that the average housing area per person in Hanoi in 1981 was still only about 3.5 square metres. In many households, that figure was below 1.5 square metres per person. The housing construction growth rate clearly remained well below the population growth rate. At the same time, housing maintenance and renovation could improve only a small fraction of the total stock of deteriorated housing. Housing construction in Hanoi was thus an urgent need and the Party considered it a top priority.

The influx of new residents putting pressure on the city's housing stock came in large part as a result of the new regime's concentration on Hanoi. The city was the capital, so the government was headquartered there, and population growth was bolstered by the promotion into its bureaucracy of many rural residents who were rewarded for their support of the Communist anti-colonials and who had solid Communist credentials – that is, they were not urbanites who had worked for or were suspected of supporting the colonial regime. The Party focused on Hanoi as the site of new cooperatives and industrial developments as well as educational institutions, and undertook many large-scale construction projects.[3] All these required workers. In these early years of Communist rule, migration was uncontrolled and the peace and stability brought about by the end of the anti-colonial war led to a rise in the birth rate. As a consequence of Hanoi-boostering, the city's housing shortage crisis might be thought of as self-inflicted.

To relieve the demand for housing, the government sought to build. Housing construction also responded to the regime's ideological orientation, which supposed that the state would provide and control all aspects of everyday life – including work, consumption, and residence – to ensure that the life of its citizens reflected the Party's national social goals. Over the three decades from the late 1950s to the 1980s, approximately 4 million square metres of housing were produced in Hanoi, double the existing stock of 2 million square metres in 1954. But population increases diminished the new housing's impact so that the average living area per capita rose to only 6 square metres by the end of the 1980s, down, as noted earlier, from 10 square metres in the mid-fifties though almost double the 3.27 square metres at the start of the decade in 1981. The government's strategy focused on building collective housing, and while the style of the units and complexes changed over time, state-provided housing was always constructed as collective, never as single-family dwellings. The first collective complexes were one-storey brick buildings housing fifteen households per building, with shared kitchens and bathrooms in a separate construction that would service up to three residential buildings, or forty-five households. These clusters also included common areas and childcare services. Mai Động is an example of this KTT housing style, which was implemented from about 1954 to 1960. These buildings could not be produced fast enough and were not of sufficient density to make a noticeable impact on housing demands. The first multi-storey KTT to be built in Hanoi was Nguyễn Công Trứ (NCT), which used the traditional construction process and materials (brick) but was a large grouping of three- to five-storey buildings in a new high-density housing style. We document the rise and fall of NCT in this chapter, to consider the fate of Hanoi's – and Vietnam's – KTT, and the contributing factors to that fate (Christina Schwenkel discusses the fate of a showcase KTT in Vinh City in chapter 1 of this volume).

Nguyễn Công Trứ: Building the Wrong Socialist Utopia

The Nguyễn Công Trứ KTT was developed as an urgent and decisive intervention to provide housing on a massive scale for state cadre and their families living in Hanoi.[4] Renovating (which effectively meant dividing up) existent housing stock from the colonial era was too time-consuming, and the one-storey KTT model was too small-scale to meet demand. NCT was the first KTT of such scale to be built in Vietnam; many others followed rapidly. The detailed project brief – detailing number, location, and height of the proposed buildings, the proposed site location and layout, the layout and size of each building, and the proposed construction materials and technologies – was prepared by the Hanoi Department of Housing and Land Management in March 1961.[5] NCT was to be built using the building materials and technologies employed under the colonial regime: brick load-bearing walls, tiled roofs, and concrete panel floors. It was built to be sound and permanent. Only a couple of months later, however, in May/June 1961, the city approved plans for another massive KTT, Kim Liên, to be built using prefabricated materials, and from this point on all KTT were built as pre-fabs.

Nguyễn Công Trứ, meanwhile, was to be located on the only readily available large parcel of land with accessible roads and infrastructure, the French cemetery An-Do. The cemetery had only been moved to this site in the 1930s (Logan 2000; Drummond 2007), taking up several hectares of land belonging to the Trưng Sisters Pagoda amid much controversy. The Committee of City Development was assigned management of the NCT project and was responsible for the relocation of graves, land levelling and drainage, supervising the project, and ensuring compliance on the adjoining sites, all in accordance with the approved city master plan. Water drainage received particular attention, it seems, for although there were no building plans available in the archives, there was detailed documentation regarding the planning and design of the drainage system for NCT. Hanoi is a city built below sea level and on swamps; it is protected from river flooding by an extensive dike system begun almost 1,000 years ago, but is frequently subjected to street flooding in the rainy season. As the first and flagship KTT, NCT's drainage needs were carefully accounted for in the plans, but many KTT built afterwards had flooding issues because of a lack of a proper drainage system.

Construction of the NCT complex took approximately two and half years, and was completed at the end of 1963 (Đặng Thái Hoàng 1985, 37–9). The finished KTT was made up of fourteen four-storey collective apartment blocks. Each housing block had four floors, and each floor was divided into two sections, with one common stairway providing access to the four floor levels. Each floor housed between sixteen and eighteen families and had a total of eight common areas for cooking and bathing. Thus on each floor there was a common area on each side of the stairwell, running along one side of an internal corridor, in which were a kitchen, bathroom, and laundry shared by the eight or nine living/sleeping rooms located across the corridor. The living/sleeping rooms were between twenty and twenty-five square metres and were intended to be used by one family based on the standard of four square metres per person. Each block housed between sixty-four and seventy-two households.

Compared to earlier socialist housing developments – Mai Động, noted earlier, and An Dương, Phúc Xá, and Mai Hương – NCT's design layout represented an improvement in terms of comfort and privacy. The earlier collective housing design located the kitchen and sanitary facilities in a separate building in the middle of the housing units. In NCT, by contrast, there were some eight separate shared kitchens and bathrooms, two on each of the four storeys, each for the use of eight to nine households. The facilities were thus geographically closer and shared among fewer households, though still shared. (In a few units, destined for higher-ranking-cadre households, the wet facilities were shared between only two to four households.) A significant feature of NCT and other KTT complexes, particularly from the early era of KTT construction, is the often quite generous common space provided between buildings. These common spaces had never been a feature of pre-existing urban architecture or design, either feudal or colonial. As such, these common spaces – meant to provide both breeze access to the housing units in the residential buildings and recreational space for residents – represented a new type of space in Hanoi. They were neither fully private spaces, since they were not under the control of any individual, nor fully public spaces, because they were located within the KTT and meant for the use of residents only; they were not open to the wider urban public.

While Nguyễn Công Trứ was originally designed simply as a high-density solution to the pressing demands for housing supply, by 1963 Vietnamese officials had begun to discuss the design principles of the Soviet concept of the mikroraion.[6] A mikroraion was an urban residential complex of a certain density achieved through multi-storey apartment buildings in which basic services and amenities were also provided to simplify daily life: a "self-contained neighbourhood unit" (Gentile and Sjoberg 2006, 706). As a result, NCT added a canteen, primary school, and kindergarten, all constructed after the residential towers in 1963–4, to retrofit the complex into a mikroraion. The next KTT to be constructed after NCT, Kim Liên and Giảng Võ, were the first to be designed from the start according to the mikroraion principles.

According to the mikroraion concept, a geographic formula dictated such things as ideal territorial size and densities, and also distances, for example between entrances to the mikroraion and between residential units and the various services – schools, medical clinics, or shops. Public transportation and roads were to be placed around the periphery of the district, preserving its internal spaces for greenery, walking paths, and amenities/service buildings. However, in practice, as Gentile and Sjoberg (2006, 706) point out, services often could not keep up with construction of the residential units and many mikroraion received their services and amenities only years after residents moved in – if ever. This was the fate of such services in most of Hanoi's KTT, many of which also lacked the basic access to public transportation that the mikroraion principles specified; instead of the rich network of public transportation envisioned for mikroraion, Hanoians rode bicycles on a road network not greatly expanded or improved since independence. Built as designed, however, the mikroraion embodied socialist living on an everyday basis: collective living for a collective society. The services within walking distance that made each neigh-bourhood "self-contained" were intended to model a perfected every-day life where residents found what they needed at a geographic scale that economized on reproductive tasks such as cooking and childcare, through the on-site provision of canteens and daycares, to make more time available for productive work. The mikroraion represented the pursuit of social goals through residential design.

Nguyễn Công Trứ KTT, once it was built and residents began mov-ing in, was perhaps the most prestigious and desirable housing allo-cation for most Hanoians (Trịnh Duy Luân, cited in Logan 2000; the highest-level Party elite were assigned official residences in some of the French-built villas kept aside for the purpose). Nguyễn Công Trứ was new. It had plumbing located within the building. It had services within the complex. It had green space immediately adjacent to each residential building. No other housing available in Hanoi was as mod-ern. Many state cadre had been in housing units that were simply rooms in divided-up existing housing, either colonial-era villas or tube houses in the crowded Old Quarter, with very basic external kitchens and sanitary facilities shared between a multitude of families. In the Old Quarter, many houses had no access to running water into the 1990s, and families still collected water from standpipes in the streets (Drummond 2000). NCT offered a welcome escape from these condi-tions, for those lucky enough to be allocated a unit there.

The KTT opened to residents in the late summer of 1963. Almost immediately, problems were reported by and among residents. The small living area allocated to each household and the shared kitchens and bathrooms created tensions among residents over how these were used and kept. Residents complained that others failed to keep the shared areas clean, tidy, and quiet. Balconies and corridors were used for drying laundry. Some residents used their rooms as workshops, and some raised pigs in them.[7] A photo of a NCT block with laundry draped outside windows on the narrow balconies was published in the *Thủ Đô Hà Nội* newspaper in November 1963, immediately after the first residents moved in, with the caption: "The house is beautiful, but do [we] have to live in it this way?" The article it accompanies details a

3.1 Nguyễn Công Trứ collective housing blocks in the 1960s – code B.5944. Copyright Vietnam News Agency.

journalist's visit to NCT and discussions with residents about the experience of living in Hanoi's first high-density collective housing:

> As I [the journalist] entered the KTT, the thing that drew my attention the most was people hanging their laundry all along the corridors. Furniture was chaotically all over the place. There were many hand prints on building walls. Chips appeared in several places in the stairwell. Water was running heavily down to the drain but still could not take away all fish scales or vegetable peels.
>
> I knocked on the door of a stranger and went in. After talking for a while, the owner took me to see the urinals. Thinking I was a cadre of the Housing Department, he asked for the door to his room facing the bathroom to be relocated. In fact, this door was facing the urinals. There was somehow an unpleasant smell spreading in his room ...
>
> I asked: Every family here has to take part in cleaning the bathroom, right?
>
> – Yes, but there are people who do it well and people who do it badly. But I always do it the most carefully.

The kitchen, the toilet, the shower, the urinals are close together. Just one person not cleaning well enough can make the bad smell spread through the whole area. If there was a clear cleaning schedule the residents would immediately know who did the cleaning badly – I suggested this to the "poor owner." (*Thủ Đô Hà Nội* 1963)

About ten years later, a journalist for the paper *Hà Nội Mới* wrote about his/her visit to NCT in much the same way:

At the KTT a month ago, there were chaotic and unpleasant views. The fourteen land lots in between fourteen housing blocks were filled with chicken cages and vegetable gardens, all placed without any order. In fact, many KTT residents – especially those living in the ground and first floors – without acknowledging each other, just occupied a piece of land for agricultural production or for expanding their living areas.

Some people even removed the stone benches, which are rare toys for children, just to expand the areas that they illegally occupied. Even knowing that it was illegal, that it was against city regulations, they still did it. That is because they hear some people expressing opinions like "don't do that" or "you should not do that" – but the authorities took no real measures. And when one person does something, other people also think that they can do similar things! Many people have been doing these wrong things, which makes them not wrong anymore! Thus they started erecting buildings, at first bicycle sheds, then kitchens, then habitable rooms, and even workshops! Like mushrooms growing under a tree trunk, fifty-five roofed buildings appeared around the fourteen housing blocks. The formidable KTT has been given the colours of the countryside with fruit and vegetable gardens, chicken farms, and chaotic buildings constructed spontaneously with scrap or human-made materials. (*Hà Nội Mới* 1975, all translations by Nguyễn Thanh Bình)

Management of the NCT KTT was assigned to the local authorities, who were given detailed requirements including the frequency of maintenance and repair, housing rules and regulations, and various issues of security, hygiene, and welfare.[8] It soon became apparent that the local authorities made little attempt to manage NCT as intended. The 1975 article in *Hà Nội Mới* here provides evidence that even then, encroachments into common spaces were frequent. The article notes that those who committed these encroachments, or enclosures, were aware that such acts were illegal but since they received only "friendly reminders" of the regulations they were infringing, they were not deterred. Such encroachments expanded household living space and provided larger living rooms or kitchens for those on the ground floor, or workshops and auxiliary space for those in units on higher floors, such as for small sheds for storing the family's bicycle so that it would not have to be wheeled up several flights of stairs. Encroachments also appropriated common land for growing vegetables and keeping chickens, important supplementary sources of food in the Subsidy Era, when even one's ration entitlement, already a very small amount of food, was sometimes not procurable.

Eventually it was the NCT Board of Resident Representatives, not the authorities, that intervened in 1975 to remove illegal constructions occupying common areas. The board held several meetings with residents to discuss the issue and persuade the builders to remove their illegal structures. Many residents accepted the direction to remove their encroachments seemingly because the structures themselves were relatively unimportant – for example, the bicycle sheds or chicken coops. It appears, however, that the board's motivation came not from an internal desire to enforce proper management and respect for the common areas, but primarily from a state directive to all units to clean up the city in preparation for Independence Day celebrations. Similarly, transgressing residents may have realized that the event was sufficiently significant that the authorities would eventually have intervened had the matter not been resolved internally by removing the offending structures. No doubt many simply intended to reinstitute their structures once the more powerful external eye of authority turned to other matters than the urban landscape and the aesthetics of the city's flagship socialist residential complex.

In other words, from the moment that the complex opened and the first residents moved in, there were clear signs that this model of socialist living created tensions and dissatisfaction. Even when "wet" facilities (kitchen, bathroom, laundry) were brought inside and onto the same floor as living/sleeping units, the collective residence was not an easy fit with tenants' preferences. Enclosure of common areas was a well-established problem within ten years of NCT's opening, and this phenomenon in particular only worsened, first with the economic crisis of the early 1980s, and then in the new conditions created by Đổi mới, the open-door policy responding to that crisis. It could be said that the flagship collective residential complex, Nguyễn Công Trứ, in its designed state and used according to its designed intent, lasted only a few years at best. This was not the residential utopia its tenants desired. And the socialist nation it was intended to help create was meanwhile beset by problems, including war and the rebuilding required after reunification in 1975, as well as unsuccessful experiments in collectivized agriculture and production.

The economic crisis of the first half of the 1980s brought new pressure to bear on the common areas of the NCT complex. The city's major "black market" operated adjacent to NCT and as it expanded – being virtually the only place where Hanoians could buy and sell goods, particularly goods informally (illicitly) brought into the country – NCT was a prime location. Many ground-floor residents extended their units into the common area between buildings and used the new space created as a shop or storage area for goods to be sold in the black market. Đổi mới – the period after the signing of the 1986 policy to open Vietnam's doors to foreign investment, allow a private sector to develop, and end the subsidy/rationing system – brought significant changes and lessened the black market's importance. However, the market itself remained and flourished, mainly specializing in electric goods and lighting, but selling many other goods as well. A 1989 newspaper article described NCT's common spaces as experiencing "a war" over land: "the appropriation of public land for illegal housing construction has become a 'real war' for land" [lấn chiếm đất công xây nhà trái phép cuộc "chiến tranh thực sự" về đất đai] (Hà Nội Mới 1989). Ground floor residents built out as much as they could, while residents of higher floors, and even nonresidents, vied to grab unoccupied sections of common areas between the blocks and built structures. Often the would-be squatters hired extra labourers to finish the work quickly, even overnight, to ensure construction would not be interrupted in process; a completed occupied structure was considered to be less vulnerable to objections or even a removal order (Hà Nội Mới 1989).

Some state organizations were allowed to develop infill housing on the common areas of the KTT. Infill development in KTTs was common and in many cases, state organizations were allowed to develop housing for their employees through infill construction. In the case of NCT, a report on the proposed redevelopment project noted the following infill construction: two apartment buildings that had been built in 1992 by the Real Estate Development and Trading Company (it is not clear if this was a private or state-owned company) and the apartments sold to private buyers; infill blocks A1, A3, A5, A7, and A9 were built in 1987 (no note of which state organization built them); infill apartment blocks constructed by the Post Office and the former Trading Cooperative [Hợp tác xã mua bán]; and infill houses around the former cemetery building (now the Ward Table Tennis Club). By the early 2000s, the total population living in NCT's fourteen collective housing blocks had reached 7,000, nearly twice the design capacity, while another 2,000 people lived in the infill housing or occupied what had been common land (Tổng Công ty Đầu tư Phát Triển Nhà Hà Nội, 2007).

3.2 Nguyễn Công Trứ collective housing blocks today. Photograph by Daniel Đặng.

Changing Tracks: Market Socialism's Urban Vision

Hanoi continued to build KTT until the early 1990s, some years after the introduction of Đổi mới, although the last complex built with central state funding was Thanh Xuân in the early 1980s; other KTT housing was constructed by work units. Altogether, 1,200 KTT were built in Hanoi, which received the majority of the nation's total of 1,500 KTT. The total population housed in KTT in Hanoi amounted to about 27,500 households or 137,000 residents. The inconvenience of the shared facilities and their hindrance to satisfactory collective living was officially recognized quite quickly, and the number of apartments sharing facilities was reduced as early as the 1965–70 KTT designs. From 1975 on, newly built KTT were designed to have all self-contained apartments. Residents of shared-facility KTT began to invest in alterations to provide private facilities as soon as they could. At the same time, however, from the late 1980s and as a result of Đổi mới, semi-private developers (state enterprises acting as private companies or state-private joint ventures) began to enter the housing construction industry.

Đổi mới produced a national socio-economic system the Party now calls "market socialism" [Kinh tế thị trường định hướng xã hội chủ nghĩa]. The policy, adopted in 1986, gradually eradicated subsidies on key goods and the ration coupon system that distributed them, and embraced some free-market principles including the development of a private sector and private ownership of property. Politically, the nation continues as a one-party Communist state. The policy has had an enormous impact on every-day life for Vietnamese, emancipating many aspects of daily life from state control and transforming many previously state-provided services into user-pay services, such as education, health care, and, importantly for us, housing. In the early 1990s, a new law paved the way for residents to be given title to the land or unit they occupied at the time, which could from that point therefore be bought, sold, mortgaged, or inherited.

Across the city region, Hanoi's urban landscape has been transformed by Đổi mới so that the central (old) city is now ringed by high-rises built in the suburbs. This period has seen a massive suburbanization of the city as adjacent villages and agricultural land were swallowed up in urban extension and, often, cleared by the state for housing devel-opment projects, with the state providing much of the compensation to landowners, generating an enormous amount of controversy and dispute.[9] Official recognition of the suburbanizing impulse came with 2009's administrative extension of Hanoi to include an entire former province and districts from other contiguous provinces. The suburbs and peri-urban areas have been the main locus of development because the small established urban core is already overcrowded and has lim-ited infrastructure for transportation, water, and so on. But the four inner districts remain the heart of the city, containing most of its eco-nomic, social, cultural, and educational services and institutions. Inner-city land is often ten times more expensive than land in the suburbs, and large parcels of land that can be allocated to a large new develop-ment are rare. The state has sold or redeveloped (often in a state-private enterprise joint-venture arrangement) many of the large land parcels that once housed state agencies or enterprises. These have been used for high-rise luxury office, hotel, or apartment buildings. The state, in the form of the municipal authorities, has even allowed developers into the city's parks and other public spaces (the most notorious and, seem-ingly, unsuccessful such arrangement being the hotel development slated for the city's largest inner park, Thống Nhất, which was scuttled by protests; see Wells-Dang 2012, chapter 5; Drummond 2007).

As the supply of these "available" lands ran out, the city authorities were, perhaps serendipitously, turning their attention to the conditions of the city's KTT. At that timely moment, the Hanoi People's Committee abruptly issued Resolution 07/2005/NQ-HDND, which stated that KTT in Hanoi would be demolished and redeveloped because they were no longer safe or suitable for residents. Demolition of the KTT for the development of new apartments for commercial sale was thus tabled as one of the city's major tasks for the period 2006–15, by which time all the KTT should have been demolished and the building of new condominiums well underway. It is not clear what constituted the basis for this state assessment of the KTT. No comprehensive or selective survey of KTT conditions appears to have been implemented, and none was referenced in the decision, nor has any survey about the experiences of KTT residents been conducted to ascertain their views on the viability of the complexes in which they live.

The resolution on KTT redevelopment proposed that redevelopments would be financed exclusively by private investment. In practice, however, the city has ended up heavily subsidizing the few redevelopments that have happened or are underway, including that of Nguyễn Công Trứ, perhaps the best-located KTT in the city and thus the redevelopment most likely to attract private investment. In part, perhaps large part, this subsidization has been necessitated by the compensation demands associated with displacing KTT residents. An important aspect of the process of conferring land title that occurred in the 1990s, and which is having significant implications for implementing the municipal authorities' current urban vision, is that when tenants of KTTs received their land titles, the documents set no conditions regarding how long the title holders might reside in that KTT or in what circumstances title holders might be forced to accept relocation or compensation to allow for the KTT to be redeveloped. This situation has become a significant impediment to redevelopment. KTT redevelopers must negotiate with existing residents who are entitled to claim a unit in the new building, though they must pay market rates for the space exceeding the square meterage of their original unit. If residents do not opt to take up a unit in the new building then they must agree on a compensation rate with the developer, and many residents, eyeing the future profits of developers, insist on robust compensation. The developer's profits are to come from building more units than the original KTT contained and selling those at market rates. Developers have accordingly proposed new buildings at least three times higher than the existing KTT blocks to realize this profit. Meanwhile the city's official master plan (to 2030) prohibits the increase of population or building densities in all KTT,[10] although to allow the KTT redevelopment project to progress the city has had to allow overbuilding above limits set by the city's regulations and official master plan.

Another aspect of the compensation issue is related to the problem of transportation infrastructure demanded by the redevelopments. Most KTT already suffer from lack of urban infrastructure such as transportation, in both the lack of public transit and the lack of space for private transportation as well as the lack of roads and street width and lack of space for parking. The road surface required to adequately service a larger population than the existing KTT would be exorbitantly expensive and require a long time-consuming process of negotiation over compensation with existing residents or shop owners on the required land. How this will be resolved is yet to be seen. Transportation, at least, comprises infrastructure deemed to be important by urban officials. Meanwhile, the city as a whole lacks public space, and the KTT have lost most if not all of theirs to encroachment and squatting. Provision of public space has not been a notable feature of the proposals for KTT redevelopment.

As a result of these issues, the sweeping program to redevelop all KTT by 2015 has progressed extremely slowly and was still far from complete in 2018. Most of the attempts to interest investors and initiate demolition and redevelopment have been unsuccessful, both because of demands from residents and the city's urban planning constraints. The very few KTT redevelopments that have been undertaken have occurred either with planning exemptions (e.g., extra height or density allowances) or subsidies from the city authorities, or both. In the case of the Nguyễn Công Trứ KTT, the city has had to not only repeatedly repeal planning restrictions to allow denser and higher towers, but has also had to subsidize about VND 300 billion of the project's total investment of VND 503 billion to settle the compensation demands of only 300 of the complex's title-holding households.[11] If the redevelopment plan proceeds as approved, the new NCT population may exceed 14,000 people, with much reduced allocation of green spaces and preschool provision compared with the redevelopment plan originally approved. There were three development options, one of which was approved, and an initial phase of demolition and new construction of one tower has begun. The total proposed apartment housing area is to be 216,650 square metres (20 square metres/person), equal to about 10,832 apartments, with a total service/commercial area of 128,025 square metres, and one kindergarten with a floor area of 2,430 square metres (built two or three storeys high).

There has been an indication that the Ministry of Construction will issue a decree that exempts new apartment buildings replacing KTT from Hanoi's planning restrictions. The proposed decree does not address the impact such an exemption will place on existing social and

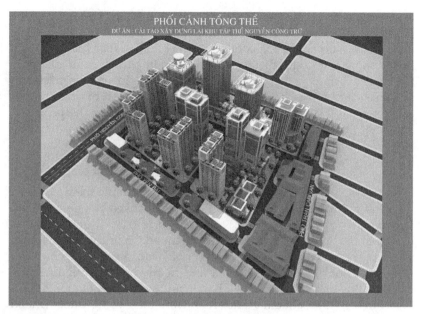

3.3 The redevelopment design concept for KTT Nguyễn Công Trứ (Option 3, approved). The low-rise buildings surrounding the high-rise buildings are actually infill buildings developed since the late 1970s. Because of these houses' high commercial value, the owners living in these buildings have strongly resisted the redevelopment program. As a result, the developer could include only old KTT blocks in the project. Image source: https://dothi.net/chinh-sach-quy-hoach/khu-tap-the-nguyen-cong-tru-boc-tham -can-ho-tai-dinh-cu-ar3506.htm.

technical infrastructure, its influence on the long-term urban vision for the city, nor the rights and concerns of residents of the communities adjacent to the KTT facing redevelopment.[12]

Conclusion: Hanoi's Future Rests between Decision and Vision

These days the issue of KTT redevelopment is understood to be complex and difficult, but there has been no questioning of the underlying principle that the KTTs should be demolished. In this sense, the decision to demolish the KTT resembles the original decision to build them: both were adopted without debate about alternatives or public input. Despite instances where residents have refused to move out of KTT proposed for redevelopment there has been little effort to understand why or research to assess residents' needs, preferences, or openness to alternative options. Nor has there been, apparently, any research into the implications of such a massive urban policy for the city as a whole. This is especially surprising considering the fact that, in many respects, the policy appears to contradict or work against important principles only recently enshrined in the master plan. Such sweeping policy enacted in the absence of research echoes the lack of assessment of urban implications when these massive KTT complexes were originally proposed and built. Just as the KTT construction model was adopted unilaterally, so now the demolition of KTT is being unilaterally imposed. In the former case, however, it is clear that the motivation was a need to solve a crisis in housing. In the demolition case, the motivation is more opaque. Although the KTT house thousands of people there has been no vocal public demand for the KTT to be removed. While some are indisputably in poor condition, Nguyễn Công Trứ, for example, was built soundly and should require only repair and maintenance to return to a reasonable condition. NCT was built with the same technology used for colonial-era construction, and many of those buildings are still standing and still in use; NCT was built with the intention of lasting "forever." Subsequent Hanoi KTT were built using prefabricated concrete slabs, but even for these constructions the expected lifespan, though shorter than for the NCT technology, would be approximately 100 years, although dependent on regular maintenance and repair.

As Christina Schwenkel (2012) has argued regarding another flagship collective housing complex, in Vinh City, the KTT have tumbled from desirable to despised in only a generation or two. In our interviews with various planners and architects, of both the pre- and post-Đổi mới generations, none indicated concern to preserve any aspect of the KTT either for practical or heritage reasons.[13] Several described the collective design of Hanoi's KTT as "inappropriate" for Vietnamese residents, and as having been "imposed" by Vietnam's socialist siblings, who contributed to their construction financially or through the provision of expertise in planning, architecture, or urban design (see Logan 1995).

While it appears that financial realities have interfered with the plan for KTT eradication, the reprieve may be short-lived because it represents a financial constraint, not a policy reversal. The urban vision that produced the KTT saw a socialist solution to a crisis of housing. Collective housing and the perfected everyday life of the mikroraion would be a lived experience in preparation for a fully collectivized society. Market socialism, however, requires a society of largely self-reliant citizens, and its urban vision confronts a crisis of housing repair to which the solution appears in the form of private developers. In both cases, the proposed solutions involve substantive, even massive, interventions in the urban landscape and everyday lives of urban residents with little study or even preparation for those impacts. The contemporary fate of the KTT shows that one era's solution can too easily become another era's burden. Full-scale conversion of the KTT into much higher density condominiums will undoubtedly produce a burden to be resolved for the next generation of urban visionaries.

ACKNOWLEDGMENTS

The authors acknowledge the Social Sciences and Humanities Research Council of Canada, which provided financial support for research, and our research collaborators, Douglas Young and Markus Kip.

NOTES

1 There is some difficulty in ascertaining Hanoi's pre- and post-Independence population figures. In general, estimates for the pre-1954 population are around 150,000. Dr Đào Ngọc Nghiêm (a former Chief Architect of Hanoi) estimated the population in inner Hanoi to have been 160,000 in 1954 (Đào Ngọc Nghiêm, personal communication). Turley (1975) uses the figure 129,461 for the urban population in 1953, but the source of this figure is not clear. Post-1954 estimates vary more widely. One source, the "Hanoi Portal" of the Hanoi municipal government, gives a figure as low as 53,000 for the 1954 population (see Hanoi Portal 2009). This seems too low. Trịnh Hồng Triển (1984) offered the figure of 370,000 in an article published in the journal *Kiến Trúc* [*Architecture*], which is much higher than other estimates. Trương Tùng (1984), then vice president of the Hanoi People's Committee, gave a figure of 270,000 in another article in the same journal issue, which may also be a little high. Dr Đào Ngọc Nghiêm (personal communication), whose pre-1954 estimate is noted here, suggests the city population was 463,000 in 1961, very close to the figure of 458,090 we use from the *Thủ Đô Hà Nội* newspaper. Turley (1975) cites 458,000 as the 1960 inner city population given by the census. We have settled on a rough estimate of 160,000 in 1954 as a reasonable reflection of the urban population of Hanoi at the moment of commencing on its socialist path and 458,090 in 1961.
2 This was an official report read by Mr Bùi Quang Tạo, as Minister of Architecture, to the National Assembly in support of the decision to expand Hanoi's administrative boundary. According to this report, Hanoi's inner city had 458,090 people and 2,795 ha (27.95 square kilometres) of land in 1961.

3 Document No.84/UB dated 24 May 1963, prepared by Hanoi Administration Committee, Hanoi Archives, Box 185, Doc. Nr. 4186.

4 "Construction Documentation for the Construction of Nguyễn Công Trứ Housing Complex," Box 141, Document Nr. 3157, 1961. Document #18/ND, 5 January 1961, issued by Hanoi Housing and Land Management Department, sent to Hanoi Department of Architecture on designing housing in the French cemetery [Văn bản số 18 ND ngày 05/01/61 của Sở Quản lý nhà đất gửi Sở KTHN về nhiệm vụ thiết kế nhà ở nghĩa địa Pháp].

5 Document #18/ND (5 January 1961) and 138 (28 February 1961) issued by Hanoi Housing and Land Management Department, sent to Hanoi Department of Architecture on designing housing in the French cemetery [Văn bản số 18 ND ngày 05/01/61 của Sở Quản lý nhà đất gửi Sở KTHN về nhiệm vụ thiết kế nhà ở nghĩa địa Pháp]. The construction of NCT was funded in total by the National Planning Committee. There is a detailed project brief in the archives, prepared by the Hanoi Department of Housing and Land Management, which sets the number, location, and height of the proposed buildings, the proposed site location and size, the layout and size of each building, and the proposed use and construction materials and technologies. But we found no record of any discussions regarding the development of NCT that resulted in these decisions. We also found no design documentation; the project brief was used by the Hanoi Department of Architecture – the city's construction department at that time – to calculate expenses and issue construction work instructions.

6 Document No. 575 UB / KT, 9 July 1963, Memo from the State Basic Construction Committee sent to the Hanoi Administration Committee on re-studying the plans for building microraions [Công văn số 575 UB/KT ngày 8 tháng 7 năm 1963 của UBKTCBNN gửi UBHCHN về việc nghiên cứu lại kế hoạch xây dựng các tiểu khu nhà ở tập trung], signed by Deputy Head Trần Đại Nghĩa (2–3).

7 Later, this was a fairly common, or at least not unusual, use of KTT balconies when food became an issue of urgency. Urbanites' access to food during the Subsidy Era (1960s to 1986) was controlled through ration coupons redeemed at state food-distribution stores, where supplies were limited if not lacking altogether. The Vietnam Museum of Ethnology 2006 exhibition on the Subsidy Era detailed these practices.

8 Hanoi Archives, Box 187, Doc Nr. 4238, 1963.

9 Recent media reports claim that up to 98 per cent of legal appeals to the government between 2015 and 2018 – a total of 21,274 cases –were related to land issues. "Khiếu nại liên quan đất đai vẫn cao" [The Rate of Land Appeals is Still High], https://vietnamnet.vn/vn/bat-dong-san/du-an/khieu-nai-trong-linh-vuc-dat-dai-chiem-98-553657.html; another media source, *The Vietnam News*, gives the figure 96 per cent in the same period:

https://www.rfa.org/vietnamese/news/vietnamnews/complaints
-related-to-land-disputes-accounted-for-nearly-96-07112019082738.html
10 The master plan was approved in 2012. Its official title is *General Construction Planning of Hanoi to 2030 and Vision to 2050*.
11 According to a redevelopment project report, the total population of NCT was 9,300 people or about 2,050 households. The total population requiring resettlement was about 8,500 people or 1,843 households. Of that total, about 7,000 people (or about 1,393 households) lived in the original fourteen KTT buildings; about 550 people (about 130 households) lived in privately owned houses (legal infill housing); and about 950 people (about 320 household) lived in illegal infill housing in the KTT complex (Tổng Công ty Đầu tư Phát Triển Nhà Hà Nội - Công ty ĐTXDPT Nhà Số 7 Hà Nội [Hanoi Housing Investment and Development Group – Housing Investment and Development Company No.7], 2007.
12 There are many disagreements with the newly proposed decree, see for example Khanh (2015).
13 We met in 2013 with architects and planners from the Hanoi Architecture Department, a former chief architect, a senior landscape architect with the Hanoi Park Company, the Vietnam Construction Association, the Vietnam Architecture Association, and the Vietnam Urban Planning Development Association.

REFERENCES

Bùi Quang Tạo. 1961. "Mở rộng địa giới Hà Nội" [Expanding Hanoi's Administrative Boundary]. *Thủ Đô Hà Nội* 773, 14 April, 1, 4.
Đặng Thái Hoàng. 1985. *Kiến trúc Hà Nội thế kỷ XIX và thế kỷ XX* [Nineteenth- and Twentieth-Century Hanoi Architecture]. Hanoi, Vietnam: Nhà Xuât bản Hà Nội [Hanoi Publishing House]
Drummond, Lisa B. Welch. 2000. "Street Scenes: Practices of Public and Private Space in Urban Vietnam." *Urban Studies* 37, no. 12: 2377–91.
– 2007. "Contesting Landscapes: Negotiating Hanoi's Urban Development under Colonial and *Doi Moi* Municipal Governments." Paper presented at the International Convention of Asia Scholars, University Sains Malaysia, Kuala Lumpur, 2–5 August.
Gentile, Michael, and Örjan Sjöberg. 2006. "Intra-Urban Landscapes of Priority: The Soviet Legacy." *Europe-Asia Studies* 58, no. 5: 701–29.
Hà Nội Mới. 1975. "Việc làm được ở khu tập thể Nguyễn Công Trứ" [Good Works Done at KTT Nguyễn Công Trứ]. *Hà Nội Mới*, 1 October.
– 1989. "Tiếng kêu cứu từ các khu tập thể" [A Scream for Help from the KTTs]. *Hà Nội Mới*, 10 March. Readers' Letters section (author's name illegible).

Hanoi Portal. 2009. "Dân số và diện tích" [Population and Territory]. http://
hanoi.gov.vn/diachihanoi/-/hn/Qnq55IvLejjp/1102/30229/dan-so-va
-dien-tich.html. eref>

Logan, William S. 1995. "Russians on the Red River: The Soviet Impact on
Hanoi's Townscape, 1955–90." *Europe-Asia Studies* 47, no. 3: 443–68.

– 2000. *Hanoi: Biography of a City*. Sydney, Australia: University of New South
Wales Press.

Schwenkel, Christina. 2012. "Civilizing the City: Socialist Ruins and Urban
Renewal in Central Vietnam." *Positions* 20, no. 2: 437–70.

Thủ Đô Hà Nội. 1963. "Chuyện nhà, chuyện cửa" [The Home Stories]. *Thủ Đô
Hà Nội*, 3 November.

Tổng Công ty Đầu tư Phát Triển Nhà Hà Nội-Công ty ĐTXDPT Nhà Số 7
Hà Nội [Hanoi Housing Investment and Development Group-Housing
Investment and Development Company No.7]. 2007. "Báo cáo dự kiến giải
pháp thực hiện dự án Nguyễn Công Trứ" [Implementation Solutions Report
for the Project Nguyễn Công Trứ] (unpublished report). November 2007.

Trịnh Hồng Triển. 1984."Hà Nội xây dựng nhà ở" [Hanoi Constructs
Housing]. *Kiến Trúc* [*Architecture*], no. 3: 21–4

Trương Tùng. 1984. "30 năm xây dựng Hà Nội [30 Years Building Hanoi]."
Kiến Trúc [*Architecture*], no. 3: 18–21.

Tuệ Khanh. 2015. "'Thả lỏng' nâng tầng chung cư: Dân "thổ cư" đứng ngoài
rìa?" ["Relaxing" the Height of Apartment Buildings: Excluding the Rights
of Local People?]. *VnMedia* [Online]. http://vnmedia.vn/VN/xa-hoi/
do-thi/tha-long-nang-tang-chung-cu-dan-tho-cu-dung-ngoai-ria
-96-3626185.html

Turley, William S. 1975. "Urbanization in War: Hanoi, 1946–1973." *Pacific
Affairs* 48, no. 3: 370–97. https://doi.org/10.2307/2756415

Wells-Dang, Andrew. 2012. *Civil Society Networks in China and Vietnam: Informal
Pathbreakers in Health and the Environment*. London, UK: Palgrave Macmillan.

4 Wrestling with the Soviet State: A Life History of Housing in Leningrad

THOMAS BORÉN AND MICHAEL GENTILE

Four decades ago, French and Hamilton (1979) lamented that the socialist city was the most overlooked in Western urban studies. The demise of the Soviet Union in 1991 spawned an upsurge of interest in it. Arguably, the very fact of its disappearance made it more researchable, as asphyxiating red tape and crippling restrictions on fieldwork were lifted. However, it still remains unnoticed by most agenda-setting urban theory, including by the most globally inclusive authors from the "Global North-West" (Bernt, Gentile, and Marcińczak 2015) of academia. Of course, "post-Soviet" scholarship shares part of the responsibility for this state of affairs for not having created a convincing bridge across the waters separating the realm of area studies from that of urban theoretical imagination (Sjöberg 2014). The necessary deparochialization process has only just started. By the time it runs its course, what used to be the socialist city will have evolved into a form of urbanism that could defy the theoretical void inherent in the opaque terminology of "posts" (i.e. post-socialist, post-communist, post-Soviet) that we have grown acquainted with over the past thirty years (Ferenčuhova and Gentile 2016; see also Bernt 2016; Hirt in Hirt, Ferenčuhová, and Tuvikene 2016). Even so, thousands of cities will continue to be influenced by numerous aspects of their socialist past, particularly through the living memories[1] of the Soviet Union's persistent housing crisis. The quest for an accurate and in-depth understanding of the socialist city is thus no less important today than it was immediately after the Soviet demise.

The current discursive framing of Russia's development under Putin as "Soviet Union 2.0" underscores this need: while private property is not likely to be relinquished, the re-emergence (or persistence) of Soviet-style power structures, coupled with the gradual evaporation of the institutions created under the post-Soviet period, is redrawing the governance setting within which Russian cities (and regions) are able and allowed to develop (Petrov, Lipman, and Hale 2013). To gain deeper understanding of the making of cities within the socialist system, this chapter therefore revisits the interaction between Soviet bureaucracy and Soviet citizens in relation to housing. The focus is on a Soviet housing career, or perhaps "experience": we will follow Yelena Alekseyevna's steps from a Stalinesque district to one of the many amorphous mass-housing districts built during the 1970s Brezhnev *époque*. Her account stems from a series of long, in-depth, recorded, and transcribed interviews conducted in Russian.[2] Yelena's memory, considered to be excellent by the rest of her family, was supported by a wide array of documents, receipts, and other official papers that she had saved throughout her life.

Housing Allocation under Central Planning

Investment in housing always lagged behind in the USSR (Andrusz 1984; Matthews 1989), owing to the sector's lack of direct association to the prioritized heavy industry and defence sectors. The roots of the long-lasting housing shortage lay in the inner logic and workings of central planning (Kornai 1992; Åslund 2002; Gentile and Sjöberg 2006), in the politics of sacrifice that coloured the early decades of Soviet power, and in the consequences of wartime destruction. Particularly during the Stalin years (until 1953), the actual floor space output was minimal (Sosnovy 1959; French 1995; Gentile and Sjöberg 2010), while improvements – at the expense of quality – were made during the decades that followed with the introduction of cheap prefabrication technologies and a general, albeit insufficient, policy shift towards production for consumption (Nove 1986; Alexandrova, Hamilton, and Kuznetsova 2004).

Most socialist polities never caught up with the housing standards of their capitalist competitors elsewhere in Europe, despite the great improvements made since 1957 in the USSR (when Khrushchev's mass housing production scheme was launched; Morton 1984). As a result, the situation had to be handled by means of a corruptible and inefficient state-run housing allocation system.

Because most housing in Soviet-type systems was administratively allocated, residential mobility was very low (Gachechiladze and Salukvadze 1995). The logic is simple: to get hold of an apartment, or rather "living space" as the Soviet jargon had it, anyone hoping for improved housing conditions faced two major time-consuming obstacles. The first was being admitted onto one of the many housing waiting lists – for example, there were lists for the disabled war veterans of the Great Patriotic War, for valued (cadre) workers, for the ill, and for "young specialists," as well as a "general" list for the vast majority – with some being significantly more prioritized than others (Gentile and Sjöberg 2013). The principal criterion for admission to the waiting list was the amount of living space enjoyed in one's current dwelling (Herman 1971; Alexeev 1988) – the standard requirement was an approximate (but variable) maximum of a mere 5 square metres per person, regardless of housing type or quality. This means that a household occupying 4.9 square metres of living space per person in a modern private apartment was entitled to a place on the waiting list, while a family with 5.1 square metres of living space per person in a communal apartment with shared or no facilities was not. (The kitchen, bathroom, and corridor were not included in the measurement of living space.) Then came the second obstacle: having to endure years or sometimes decades on the waiting list.

Both obstacles could have been overcome using *blat* (see Ledeneva 1998; Arnstberg and Borén 2003), a system in which power, influence, connections, tradable assets, or bribes could trump the sclerotic Soviet bureaucracy at the expense of those who were outside of the system (Morton 1980; Hamilton 1993; Domański 1997). After the entire process was completed, whether formally or informally, few people were eligible or eager to change their place of residence. Although there was a system of apartment exchanges (see Barbash 1988), it was subject to regulation by unhelpful authorities that did little to stimulate what was seen as economically rational intra-urban mobility, even from the planning agencies' point of view – they increasingly saw long commuting times as a negative influence on productivity (Sawicki 1977, 174–80). Inter-urban mobility was even more limited because of a combination of factors including the rigid housing allocation system (moving to a new city would have often meant starting one's housing career from zero, *ceteris paribus*), but especially the existence and relative enforcement of a two-sided system of migration control and planned settlement encompassing the internal passport on the one hand, and the compulsory residence registration (the notorious *propiska*) on the other (Buckley 1995; Höjdestrand 2004).

The combination of low residential mobility and migration restrictions meant that there was a significant suppressed demand for housing, particularly in central locations. With only a handful of exceptions, most notably Moscow (Light 2010) and Tashkent (Hojaqizi 2008), the socialist-era restrictions on residential mobility and migration vanished in the 1990s together with the system that created them. Households were suddenly, at least theoretically, offered the option of operating freely on the housing market to adapt or improve their living conditions. However, a range of new constraints limited the impact of the newly unleashed market forces. While time-consuming restitution policies favouring pre-socialist owners and their heirs slowed down the process of residential change in some countries (Reimann 1997; Stan 2006), arguably the greatest obstacle was on the demand side: transition-related economic hardship took a high toll on large segments of the population, postponing and altering the character of their participation in the housing market. Put differently, the housing allocation decisions and practices of favouring residential immobility under socialism continue to exert a durable and decisive impact on the social geography of the post-Soviet city (Gentile 2015), and the erosion of the urban "stiff landscapes" (Borén 2009) of socialism is taking place more slowly than expected.

To understand why, we turn to Yelena Alekseyevna's housing experience in Leningrad. We initially focus on the semi-centrally located two-room *Stalinka* (Stalin-era building) apartment in which she grew up and lived for forty years, followed by a detailed account of how she and her family were forced to move to one of the remotely located new high-rise estates. While most research on socialist housing allocation practices features the hardships involved in gaining access to housing – any housing, really – Yelena Alekseyevna's story portrays the system from the insider's perspective, adding greater nuance to the dichotomous narrative of inclusion-exclusion (from access to housing resources) that frequently emerges in the literature.

Coming to Leningrad

Yelena Alekseyevna's family came from Novgorod, where her father had a leading position after the revolution. He was a member of the ruling Communist Party and thus part of an elite that should have acted as a role model:

> But later this meant, this fact [being a Party member], that he was appointed to be the chief of the militia in some kind of *uyezd* [administrative unit]. There they had to get rid of the moonshiners, but he had been drinking with them. So they took him too. They also excluded [him] from the Party – he had troubles, and so he later travelled to Leningrad. And so he lived at the Nikol'skiy church ... there used to be a tobacco factory near the Nikol'skiy church. This is where he lived. I even remember we

went there to visit. He worked there. And later they gave him a room on Vasil'evskiy [Island].

Yelena Alekseyevna's father came to Leningrad in 1924 and could later arrange for his wife and two sons to join him. At this time the rest of the family stayed in Novgorod. In those days, it was quite common for people to live at their workplaces and even sleep in shifts in places assigned to them. It would not have been very practical to bring the entire family since they did not have a place to stay. Yelena Alekseyevna continues:

By 1927 I was already born here [in Leningrad] ... So, I was born on Vasil'evskii Island in 1927. We lived on the 4th–5th line – we had a room there[3] ... A room! In a *kommunalka* [a communal or shared apartment] on the ground floor. And then my dad was a manager [*nachal'nik*] [so] they gave him the authorization to [receive] an apartment, this [was] in 1931 ... so [we moved] to that apartment on Kievskiy street. I was four years old. And when they went to register, they said "either the whole apartment, or none at all" because there were three children, Misha, Tolia, and me. And they gave them this apartment, and we lived there for forty years.

Yelena Alekseyevna often speaks about this apartment, which they got when it was brand new, and how she did not want to move from it but was later forced to. It was located in Moskovskiy District (*rayon*) in Kievskii Street in the ring of industrial areas surrounding the historical centre. She recalls that the apartment had two rooms, nineteen and fifteen square metres respectively, and a kitchen that was twice the size of the one they have now. It did not have a bathroom, so they had to use the nearby bath houses, which she visited with the children. Both she and her husband usually washed up at their work places after or during work hours.

She underlines that they had two rooms – a great achievement compared to living in a *kommunalka*, which had low status and was on the ground floor, by many still considered to be the worst. This is not only because there is more visibility from the street and the risk for burglary is perceived as higher, but also people assumed that the moisture from the cellars made the apartments cold and unhealthy. There were also more mosquitoes at the lower levels, at least in St Petersburg, built as it is on a river delta. Also today, urban vertical space has its own hierarchy with the top and bottom floors being the least attractive (Gentile 2015).

Apart from the incident with the moonshiners, Yelena Alekseyevna's family appears to have greatly advanced its housing career in the new Soviet state. Starting from nothing at all, they moved to a room in a kommunalka and then to a two-room apartment of their own at a time of extreme and rapidly worsening housing shortages (Parkins 1953). This was far more than what most people could have hoped for until well into the Brezhnev era (Gentile and Sjöberg 2013). They were also able to take the step to a big city, establishing themselves as key people within the working-class "proletariat," which was highly valued in Soviet rhetoric, not least because of its ideological sanctity in the Marxist scriptures (e.g., in Marx and Engels's *Communist Manifesto*), but also because its input was needed for the rapid industrialization commanded by Stalin. According to Yelena Alekseyevna, her father received the apartment because of his leading position and perhaps also because he had enough confidence to demand the whole apartment for his family. Most of the apartments in the building were actually kommunalki. Evidently, socio-professional status and confidence produced the kind of "clout" that could make a difference when the issue of housing was at stake. While the working proletariat may have enjoyed the privilege of ideological glorification and the promise of untold future communist bounty, *leading* the proles was certainly a more auspicious avenue to success in the suppressed Soviet housing "market." In this sense, Yelena Alekseyevna's account resonates with the extensive literature that suggests housing was a crucial part of the communist-era reward system (Morton 1980; Szelényi 1983; Domański 1997; Gentile and Sjöberg 2013).

They subletted one of the two rooms, so the whole family – her mother and father and now three brothers and herself – lived in one of the rooms. They continued to sublet one room after the war until 1960 when Yelena Alekseyevna's second child was born.[4] Today Moskovskiy District is generally considered to be a high-status district, partly thanks to its relatively central location and partly because the apartments here are fairly large, with high ceilings, and of reasonably high quality. The main thoroughfares in Moskovskiy District are lined with showcase Stalinesque architecture, but Yelena Alekseyevna's building and the ones surrounding it were built slightly before the characteristic Stalinist "baroque" architecture was in full bloom. Therefore, they are not architecturally remarkable, and they are close to factories instead of along the main arteries of the city. Yelena Alekseyevna tells about the house and the area:

> They built it in 1931. It is very robust. And when they started to shoot from the Pulkovskiy heights [during the Second World War], our house was not [seriously] hit ... [It was] during the war, and the blockade, and in 1943 they bombed our district, they also dropped many shells toward our house, but for some reason they missed. It is a cinder-block house, but

with wooden beams. Apparently the logs were not of good quality, and a worm ate the logs. And it turned into rubbish (*trukha*) ... So we lived there for forty years and when they opened up the floor with probing instruments and pinched these boards, they were rubbish. So they did selective repairs, and where it was impossible, [where] the beams could fall, they made rails, girders.

She remembers the area:

There was a gas plant, a power station, it was such a contaminated area. There was much dirt, [and] dust. Do you understand? A bread-baking plant, soap [works], many factories, [and] we lived in that area. And cars went by night to the bread-baking plant, these bread [cars]. And this rattled the building ... We lived on the fourth floor. We had the kitchen facing the yard, and two windows facing the street, where freight trams brought [goods] to the house plant, you know, to Bodayevskie ...

And there are such huge poplars there now. There also used to be a fountain, a stage, [and] benches, and it [the house] is still standing strong, and it used to have a kind of grey colour, but now they have painted it yellow. I have friends there. They died, Vasilii Pavlovich and Shuron'ka, those blockade survivors. But their sister is still alive. She is eighty-five years old now. Well, I should visit her, but I never have the time.

Their history on Kievskii Street, where they lived since 1931, ended in April 1971, when they were forced to move. The building needed be totally renovated (*kapremont*) and those who lived there were assigned newly built apartments. Yelena Alekseyevna recounts how they tried to resist being evicted by turning down every offer, but when the case went to court they had to give up and reluctantly move to the three-room apartment they have lived in since.

From Moskovskiy to Ligovo

Yelena Alekseyevna explains how they ended up in Ligovo, a high-rise housing district in the outer southwestern parts of the city (figure 4.1). She has saved all the documents related to the move and goes through them and remembers during the interview. It all started in 1969. Within three years all the tenants had been assigned new apartments around the city. As she puts it, the *gorispolkom* (the city executive committee), the authority that decided upon many important urban issues during the Soviet era, "grabbed" their apartment block.

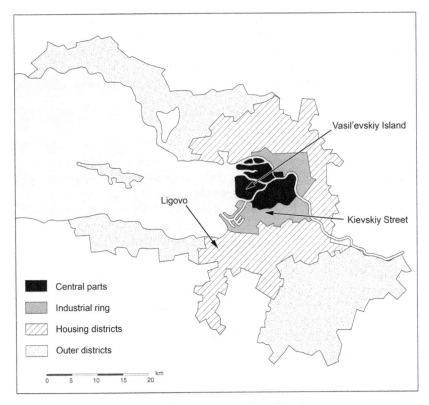

4.1 Map of Leningrad/St Petersburg, with Yelena Alekseyevna's places of residence. Map by Stefan Ene based on Ungern-Sternberg, Palčikov, and Tov 1993, 154, and Müller 1994, 147.

Because most of the apartments were *kommunalki*, one might expect that if the tenants got apartments of their own, the move was to something better, at least if one considers only the formal housing standard. But for Yelena Alekseyevna and her family, who earlier belonged to the privileged groups, the move did not feel like an improvement. Her father had passed away, and the family did not have the contacts or other means to avert the move or to gain the right to return after the renovation. Of course, from the city's and the state's perspective, by renovating the building and assigning apartments in the newly built housing districts, the population was being given something better, and more modern. The authorities were fulfilling the implicit Soviet social contract by supplying welfare in exchange for the population's acceptance of their power monopoly. In this case, an obsolete building was going to be renovated and the inhabitants were going to be given newer, better apartments and, hence, part of the welfare that the regime used to legitimize itself. To reject what was being offered was not a conceivable option.

The authorities also had plan goals to fulfil, including increasing living space (excluding kitchen, corridor, toilets, and such spaces) to 9 square metres per capita and decreasing the share of those living in kommunalki. With the family of five (now consisting of Yelena Alekseyevna and her husband, mother, and two children) in the 34-square-metre apartment on Kievskiy, this was clearly under the goal and under the mean living space both in the Soviet Union as a whole (7.4 m^2/person) and in Leningrad (8.3 m^2/ person; figures based on Morton 1973, 120 and 125.) However, in the late 1960s, as a rule, housing waiting lists in Leningrad admitted only applicant households with less than 4.5 square metres per capita (DiMaio 1974, 119). Soviet legislation offered people strong protection from eviction (Höjdestrand 2004), which was permissible only under the condition that the tenant was going to be offered something that was *at least* equivalent (DiMaio 1974, 140).

Yelena Alekseyevna took note of the number of square metres in all the apartments they looked at. If they really had to move, they wanted something bigger and better than what they already had. Apart from the number and size of the rooms, she considered which floor the apartment was on, its location, the balcony and other facilities, view, surrounding roads and possible disturbances, and not least the actual layout of the apartment. Like many veterans of the Soviet house hunt, she has great knowledge of the different types of building and apartment designs. This is because housing was built according to a few standard models, and despite there being millions of apartments throughout the Soviet Union built at the time, they all fit into a limited set of similar *tipovye proyekty* (design projects), designated by number codes that are easy to learn. Such knowledge was essential to manoeuvre in and understand the city at the time.

Yelena Alekseyevna turned down many apartments before eviction became inevitable. The first was located in Sosnovaya Poliana, not far from Ligovo, and it had three rooms. Two were connecting (*smezhnye*) and 18 and 14 square metres, and the third was separate (*izolirovannaya*) and 10 square metres (see example in figure 4.2). Separate rooms or izolirovannye are only connected to the hallway, giving the residents more privacy since they do not have to cross other rooms while moving around the apartment. Another reason is that separate rooms are easier to sublet, which makes them an important potential economic resource. This particular apartment was located on the ground floor.

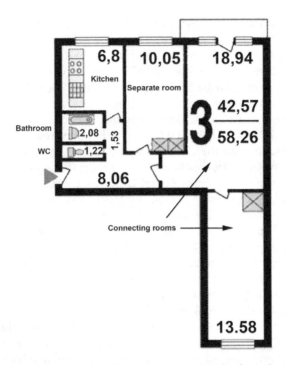

4.2 Separate and connecting rooms in a typical apartment plan from about 1970. Note also the small kitchen, and the WC separate from the bathroom, which is convenient if many people live in the apartment. The size of the rooms is indicated in square metres and both the total space and the living space of the apartment are marked, as well as the number of rooms. Figure by Stefan Ene based on http://rc.yusin.ru/uploading/big/big_9ef3cf5329t5x.jpg (accessed 16 September 2015).

Yelena Alekseyevna exclaimed "Why [should we accept] the ground floor when we had an apartment on the fourth?"[5]

A new offer came ten days later: it was in a five-storey building in Ligovo, with no elevator (Soviet legislation prescribed elevators for only buildings taller than this). She noted that it was a good apartment, but small ("39 square metres and 38 centimetres"). They declined. About two months later, in November 1970, they received an offer that they were ready to accept. She explains:

Then they gave us one far away, at Mel'nichnye ruch'i past the Finland station. This was the third display (*smotrovaya*) [apartment]. There you go. They gave it to us on the 17th of November. But my mom was in Priozërsk at the dacha and her signature was needed. While I went to her in Priozërsk, the inspector – I've forgotten her family name, maybe I will remember later – she took someone else and gave it [to them]. The apartment was in the same [type of] building as this one, a *korabl'* (ship), exactly like this. [It was] on the fourth floor, and all the rooms were izolirovannye. 18 [square metres] – izolirovannaya, 14 – izolirovnnaya, and 12. "Good" it is written … Aha, the fourth floor, a good floor, izolirovannye rooms [and] no smezhnye.

Yelena Alekseyevna then waited for her mother and when she came back after about a week they went to fill in the forms, but the inspector said that they were too late: "And again she gave us a *raspashonka* in Uritsk. And what is a raspashonka?" She points at the five-floor building just outside of the window, and says that such buildings have 44-square-metre raspashonka apartments. It had no corridor and only a single door, and she compares it with the one where they live now:

Here we at least have a corridor. Here we have a coat hanger, a shoe rack, [and] a cabinet (*penal*). But there – nothing. Can you imagine how [it is] there? … You walk into the living room, which is also considered a corridor. But there – a room – raspashonka. You walk into the living room, on the right side – a door, on the left side – a door. What kind of apartment is that? And no corridor! And the kitchen [was] a square metre smaller.

She describes the offered apartment as a "raspashonka" (lit. baby's vest) since it was small but long and narrow, like a sock from which you would enter from the middle. The apartment also had a small balcony, but the main reason they declined this one was that it was on the second floor, and the main entrance door was covered by a "screen" from which one could climb into the kitchen from outside.

The fifth offer came early in December 1970, also in Ligovo. It was on the ninth floor (the top floor) in a korabl', and was 42.5 square metres. The word korabl' means "ship" and designates a high, long, and narrow building (figure 4.3). She lists the size of the rooms and says that the elevator could have broken down, and it would have been too much for her mother to walk up the stairs. In general, top-floor apartments were not attractive since it was often colder and their flat roofs were prone to leaking. In some tipovye proyekty there were problems with the warm water not "reaching" all the way up to the top of the building. The ideal apartment should thus have been somewhere halfway up the building.

On 11 December they were offered a sixth apartment, in Kupchino in Moskovskiy District – the same district where they lived. It was on the fifth floor with three rooms and a total of 42.6 square metres of living

4.3 Korabl' or ship building. This type of high, tall, and narrow building is often referred to as a korabl', meaning ship. They are common throughout Russia and the former Soviet states. The Gulf of Finland is in the background. Photo: Thomas Borén, 2000.

space. She remembered that it was a good apartment and that it had a similar plan to the one they have now, with two adjoining rooms, one separate, and a corridor. The kitchen was bigger than the one they have today. They refused anyway: they did not want to move. Kievskiy Street was their *home*. Even when they were in dire straits, the family kept their apartment despite being offered exchanges with a lot of compensation money.[6] The apartment was the family's major asset and now the city wanted to take it away from them.

Then came the seventh apartment, in Ligovo again, but it didn't have a balcony and all of the windows were in direct sunlight, which would have "fried" them, she says. Three rooms in 39 square metres was too little an improvement. They turned down the offer once again, and by this time the apartment "search" had started to get on their nerves.

By early January, they refused the eighth apartment the authorities had come up with. Again in Ligovo, it was identical to their current one. The ninth offer came on 11 January for a place in Mel'nichnye Ruch'i. It was too far away, but she added that one could have banged on the floors as much as one would have liked because the lower neighbours were deaf.

The tenth apartment was also in Mel'nichnye Ruch'i, but in the same street and building, and identical to the previous one. The only difference was that it was on the fourth floor instead of the third. Yelena Alekseyevna saw this as pure mockery, so they did not even go to look at it though her husband was willing to accept it. He was sick of it all.

The eleventh apartment was in Ligovo, the same as one they had inspected earlier. Yelena Alekseyevna became upset, looked at the inspector and told her: "Shame on you! How can you offer us the same apartment, as here, and here? Who gives you the right to do so? Exactly the same apartment!?"

She says that "they" mock people and called the inspector a scoundrel. "We refuse – she sends us one that is the same. We refuse – she sends us one that is the same." However, by now the inspector and the gorispolkom had had enough, and decided to take them to court. According to Yelena Alekseyevna, the court claimed to understand why they did not want to move, but it nevertheless assigned them the apartment they now live in. With a new place to live, that the court and the city regarded as equal or better than what they had, there was no formal impediment to their eviction from the Kievskiy Street apartment.

Yelena Alekseyevna didn't give up. She wrote to the local department of the Party. For a moment this seemed to help. She recounts that they were offered one more dwelling, this time on Prospekt Kosmonavtov, in her "old" Moskovskiy district at Park Pobedy (a large park with a subway station). Nevertheless, they turned down the offer anyway because it was too small to also house her mother, who was afraid of being assigned living space in a kommunalka because she did not want to live with "strangers." Moreover, the balcony was small and shared with a neighbour. So they ended up in their current apartment in Ligovo, the last family to be forced away. The construction workers had already started the renovation.

In April 1971, they moved into the newly built apartment on Partizan German Street in Ligovo. By this time, Leningrad had finally managed to meet the 9-square-metre living-space target established in 1922 for the first time (Lisovskii 1983, 58; Andrusz 1984), although much of it in inferior-quality kommunalki. The system won this battle, laws and planning goals were fulfilled, and Yelena Alekseyevna and her family got the square-metre-welfare that they were entitled to, without asking or even joining the coveted waiting list.

Starting from a room in a kommunalka as a child, Yelena Alekseyevna spent much of her youth and adult life in a two-room apartment with "partial conveniences" (the Soviet way of saying there was a kitchen but no bathroom). In 1931, when they received their first apartment, and especially during the 1940s and 1950s, this would have been considered a privilege. Because Soviet housing law strongly protected tenants from eviction (which was only possible in exchange for "a bit more equal" property), this privilege was reproduced in the early 1970s. At a time when being assigned a full apartment was still unusual (Gentile and Sjöberg 2013, 181), and access to the waiting lists remained strictly limited, Yelena Alekseyevna and her family were upgraded to a full three-room apartment because they *already* had good conditions to start with. In other words, a privilege gained shortly after the revolution was reproduced forty years later, even though its value was watered down by the overall improvements in living conditions that had been taking place since the demise of Stalinism. Although Yelena Alekseyevna's expectations might have been higher, her fate is the opposite of what millions of Leningraders experienced. Decade after decade most people were forced to endure the pauper conditions that characterized life in the kommunalki of the decaying city centre.

Summing up, Yelena Alekseyevna's housing career involved three moves: all outward and upward, and to larger apartments. The outward move was from the city centre to the industrial ring and further to the *spalnye rayoni* (dormitory suburbs) that have come to symbolize the socialist city, reflecting also the relative shift from production to consumption that the Soviet Union had slowly embarked on. The upward move, from the ground floor to the fourth (of five) and to sixth (of nine) placed her and her family in a better position in the city's vertical space, on one of the mid-level *bel-étages* of the socialist city (see Gentile 2015). This did not happen by chance: new dwellings were typically produced on extensive greenfield sites that were inevitably located in the outskirts, and improved mass-production construction techniques allowed for greater vertical growth.

Meanwhile, the acute Soviet housing shortage, which had escalated dramatically during the Stalin era, had gradually eased. This favoured the slow development of a consumer society based on (more) individualistic values, while eroding its collectivist pillars, including the still ubiquitous communal living arrangements. Despite the improvements that took place during the 1970s and 1980s, communal living remained common until the very end of the Soviet period, especially in cities with large prewar housing stock. Even today, over 600,000 St Petersburg residents remain in the firm grip of the kommunalka (Pravitelstvo Sankt Peterburga 2016).

During the three decades that have passed since the dissolution of the Soviet Union, market forces have to some extent "corrected" the imbalance between demand and supply in the housing market. In the areas surrounding the larger and, increasingly, medium-sized cities, suburbanization has accelerated in intensity. In its most advanced form, around Moscow, it has even led to the emergence of a Russian brand of edge cities (Golubchikov and Phelps 2011). Meanwhile, in the city centres, gentrification is advancing, albeit at varying speeds: the process has been rapid in parts of Moscow (Badyina and Golubchikov 2005), whereas it stumbled upon a number of serious obstacles in St Petersburg, limiting its extent and ensuring a continued social mix in the inner city (Axënov 2014; Bernt 2016). In other words, while worldwide urban trends have emerged in Russia, they have thus far not resulted in the emergence of a uniform "post-Soviet" urban structure, despite the manifest unfolding of an urban governance combo based on authoritarianism flavoured by neoliberal ideology (Kalyukin, Borén, and Byerley 2015; Büdenbender and Zupan 2017). One of the reasons for these diverse urban outcomes lies in the nature of the housing privatization process of the 1990s, which played out differently in different parts of the country, leading to substantial regional variation in the share of owner-occupiers. A particularly important source of differentiation lies in the fact that privatization, which the local authorities across the country sometimes made affordable and attractive and sometimes not so much (Bater 1996), took place at the level of the individual apartments rather than building-by-building. This means that any single building may host a range of housing tenure forms (Bernt 2016), making the responsibility for its semi-private parts (staircases, elevators, etc.) unclear.

For most people, however, who like Yelena Alexeyevna and her family already had a decent place of living, the dramatic changes in the housing system would have little impact, and indeed, residential mobility in Russia remains low when compared to Western Europe or North America (Zavisca 2012). Like most people, Yelena Alexeyevna followed these changes and reflected upon what they meant for her. Yelena Alexeyevna stayed in her apartment until she died in the mid-2000s, and her family kept living in it after that.

Conclusion

The story told in this chapter is an empirical and grounded example of the interaction of life worlds and of the political system in urban Soviet Russia, contributing nuance to discussions of Soviet housing from the perspective of a "housing insider." Millions of people across the former Soviet Union and its satellite states can relate to Yelena Alekseyevna's experience, even though most never faced formal eviction. The socialist housing allocation experience, and the mythologies surrounding it (Zavisca 2012), flavour the life worlds of most post-Soviet households. In a world of square-metre norms, waiting lists, and standardized apartment designs, Soviet housing careers were central parts of the social contract between the individual and the state, and today also reflect an understanding of the common past.

Under communism, the course of urban development was shaped by a strikingly stable script aimed at creating a new and progressive city by applying socialist town-planning principles and by carving out as many elements as possible from the bourgeois past (French 1995; Borén and Gentile 2007). This effort produced a dense urban layer made of streets, houses, factories, offices, open spaces, closed spaces, monuments, and other communist paraphernalia, as well as stores and other places needed for everyday subsistence. With the demise of communism, research attention rapidly shifted towards the places, spaces, and geographical patterns produced by the transition processes, both as independent forces and in symbiosis with the idiosyncrasies of local context. However, little work has been done on the prolonged patterns of urban development of the cities in the region (Kalyukin and Kohl 2019). Urban transformation is rather the word of the day, but as Sýkora and Bouzarovski (2012, 45) argue, "the reorganization of urban landscapes in post-communist cities that began with the institutional reforms of the 1990s is far from complete." Specifically, while many formal political, administrative, and economic institutions were until recently thought to have assumed stable forms, social institutions based on the life worlds of people follow deeply rooted codes rather than formal rules, and as such they are still subject to influences from the socialist past. Understanding these past life worlds in greater depth is therefore crucial because they represent an inextricable part of the present life worlds. As Piirainen (1997) explains, the experiences of everyday life were not on the radar of Soviet social research, and until the end of the 1980s researchers were not allowed to conduct independent field studies, which imposed severe limits on the theoretical and empirical development of this area of expertise. By the time these restrictions were eased, if not entirely removed, understanding the socialist system had arguably lost some of its scholarly appeal.

Collective memory and narration within families and other personal networks constantly construct and reproduce images of the past that influence the present. How this everyday collective memory of the Soviet city forms the possibilities and restrictions of the current urban space-forming elite is a matter for further research, but it is difficult to see that waiting lists and square-metre restrictions add political value when politics in Russia today somehow romanticize the Soviet epoch. Rather, political calls for increased investments in housing, for example to erase the few remaining kommunalki, might be met with greater understanding than otherwise. However, even this cannot be taken for granted: the recent and unexpected mass protest movement in Moscow against the city's planned demolition of *Khrushchёvki* (Guardian 2017) – home to hundreds of thousands of people – shows that the appeal of improvements in the material qualities of housing is trumped by the qualities of place, and home, shaped through long-term residence (see also Trumbull 2014 for St Petersburg). Therefore, addressing Sjöberg's (2014) critique, a point that could be brought back to the field of urban studies more generally is that the study and understanding of cities located outside of the post-Soviet realm may require greater consideration of how current urban forms and life worlds develop under the influence of the city's living memory.

NOTES

1 These living memories are also kept alive by still-popular Soviet films, for example, such as *Ironiya sud'by* (1975) or illustrative clips on YouTube; see e.g. the clip from the film *Ne poedu* (1965) showing a lady who refuses to move into the newly built housing districts: https://www.youtube.com /wquHPMcU3fE?list=PL37OROGR5PDtzjbaH7SdbdQalbCg6W6wt

2 The interviews were conducted by Thomas Borén as part a long-term fieldwork in Ligovo, a high-rise suburb in St Petersburg, during sixteen months from 1998 to 2000 (see Borén 2009).

3 Some streets in the central parts of many cities that used to belong to Tsarist Russia are called "lines." Helsinki, Tallinn, and St Petersburg are examples of cities with line streets.

4 The family had been severely traumatized by the war; all three brothers died, the elder two at the front and the younger from starvation.

5 In Russian parlance the ground floor is called the first floor, meaning that the next floor is called the second floor, and not the first as in many Western cities. In this interview and throughout the rest of the text we use *ground floor* instead of *first floor* but have not changed the numbering of the other floors when mentioned.

6 This exchange-based pseudo housing market is under-researched and in-depth studies are needed to understand the role it played in major cities. See DiMaio 1974; Barbash 1988 for more on the system of apartment exchanges.

REFERENCES

Alexandrova, Anastassia, Ellen Hamilton, and Polina Kuznetsova. 2004. "Housing and Public Services in a Medium-Sized Russian City: Case Study of Tomsk." *Eurasian Geography and Economics* 45, no. 2:114–33. https://doi .org/10.2747/1538-7216.45.2.114

Alexeev, Michael. 1988. "Market vs Rationing: The Case of Soviet Housing." *The Review of Economics and Statistics* 70, no. 3: 414–20. https://doi.org /10.2307/1926779

Andrusz, Gregory D. 1984. *Housing and Urban Development in the USSR.* Albany, NY: SUNY Press.

Arnstberg, Karl-Olov, and Thomas Borén, eds. 2003. *Everyday Economy in Russia, Poland and Latvia.* Stockholm: Almqvist and Wiksell.

Åslund, Anders. 2002. *Building Capitalism.* Cambridge, UK: Cambridge University Press.

Axënov, Konstantin. 2014. "Between Degradation and Gentrification in a Post-Transformational Metropolis City Center: The Case of St. Petersburg." *Eurasian Geography and Economics* 55, no. 6: 656–73. https://doi.org/10.1080 /15387216.2015.1041540

Badyina, Anna, and Oleg Golubchikov. 2005. "Gentrification in Central Moscow – A Market Process or a Deliberate Policy? Money, Power and People in Housing Regeneration in Ostozhenka." *Geografiska Annaler: Series B, Human Geography* 87, no. 2: 113–29. https://doi.org/10.1111 /j.0435-3684.2005.00186.x

Barbash, Nataliya B. 1988. "Rayony, Kotorye my Vybiraem" [The Neighborhoods That We Choose]. In *Moskva: Gorod i Chelovek* [*Moscow: The City and the Individual*], edited by Evgeniy V. Murav'ev and Vladimir N. Peshkev. Moscow, Russia: Moskovskii Rabochii.

Bater, James H. 1996. *Russia and the Post-Soviet Scene: A Geographical Perspective.* London. Arnold.

Bernt, Matthias. 2016. "How Post-Socialist Is Gentrification? Observations in East Berlin and Saint Petersburg." *Eurasian Geography and Economics* 57, no. 4/5: 565–87. https://doi.org/10.1080/15387216.2016.1259079

Bernt, Matthias, Michael Gentile, and Szymon Marcińczak. 2015. "Gentrification in Post-Communist Countries: An Introduction." *Geografie* 120, no. 2: 104–12.

Borén, Thomas. 2009. *Meeting Places of Transformation: Urban Identity, Spatial Representations and Local Politics in Post-Soviet St Petersburg.* Stuttgart, Germany: Ibidem-Verlag.

Borén, Thomas, and Michael Gentile. 2007. "Metropolitan Processes in Post-Communist States: An Introduction." *Geografiska Annaler: Series B, Human Geography* 89, no. 2: 95–110. https://doi.org/10.1111/j.1468-0467.2007 .00242.x

Buckley, Cynthia. 1995. "The Myth of Managed Migration: Migration Control and Market in the Soviet Period." *Slavic Review* 54, no. 4: 896–916. https://doi.org/10.2307/2501398

Büdenbender, Mirjam, and Daniela Zupan. 2017. "The Evolution of Neoliberal Urbanism in Moscow, 1992–2015." *Antipode* 49, no. 2: 294–313. https://doi.org/10.1111/anti.12266

DiMaio, Alfred John. 1974. *Soviet Urban Housing*. New York, NY: Praeger.

Domański, Bolesław. 1997. *Industrial Control over the Socialist Town: Benevolence or Exploitation?* Westport, CT: Praeger.

Ferenčuhová, Slavomíra, and Michael Gentile. 2016. "Introduction: Post-Socialist Cities and Urban Theory." *Eurasian Geography and Economics* 57, no. 4/5: 483–96. https://doi.org/10.1080/15387216.2016.1270615

French, R. Antony. 1995. *Plans, Pragmatism and People*. London, UK: UCL Press.

French, R. Antony, and F.E. Ian Hamilton, eds. 1979. *The Socialist City*. Chichester, UK: John Wiley and Sons.

Gachechiladze, Revaz, and Joseph Salukvadze. 1995. "Tbilisi and Its Metropolitan Region: Social Problems in Space." In *The New Georgia: Space, Society, Politics*, edited by Revaz Gachechiladze, 153–67. London, UK: UCL Press.

Gentile, Michael. 2015. "The Post-Soviet Urban Poor and Where They Live: Khrushchev-Era Blocks, 'Bad' Areas, and the Vertical Dimension in Luhansk, Ukraine." *Annals of the Association of American Geographers* 105, no. 3: 583–603. https://doi.org/10.1080/00045608.2015.1018783

Gentile, Michael, and Örjan Sjöberg. 2006. "Intra-Urban Landscapes of Priority: The Soviet Legacy." *Europe-Asia Studies* 58, no. 5: 701–29. https://doi.org/10.1080/09668130600731268

– 2010. "Spaces of Priority: The Geography of Soviet Housing Construction in Daugavpils, Latvia." *Annals of the Association of American Geographers* 100, no. 1: 112–36. https://doi.org/10.1080/00045600903378994

– 2013. "Housing Allocation under Socialism: The Soviet Case Revisited." *Post-Soviet Affairs* 29, no. 2: 173–95. https://doi.org/10.1080/1060586x.2013.782685

Golubchikov, Oleg, and Nicholas A. Phelps. 2011. "The Political Economy of Place at the Post-Socialist Urban Periphery: Governing Growth on the Edge of Moscow." *Transactions of the Institute of British Geographers* 36, no. 3: 425–40. https://doi.org/10.1111/j.1475-5661.2011.00427.x

The Guardian. 2017. "Protests in Moscow at Plan to Tear Down Soviet-Era Housing in Affluent Areas." *The Guardian*, 14 May. https://www.theguardian.com/world/2017/may/14/people-protest-moscow-plans-to-tear-down-housing

Hamilton, Ellen. 1993. "Social Areas under State Socialism: The Case of Moscow." In *Beyond Sovietology – Essays in Politics and History*, edited by Susan Gross Solomon, 192–225. Armonk, NY: M.E. Sharpe.

Herman, Leon M. 1971. "Urbanization and New Housing Construction in the Soviet Union." *American Journal of Economics and Sociology* 30, no. 2: 203–20. https://doi.org/10.1111/j.1536-7150.1971.tb02959.x

Hirt, Sonia, Slavomíra Ferenčuhová, and Tauri Tuvikene. 2016. "Conceptual Forum: the 'Post-Socialist' City." *Eurasian Geography and Economics* 57, no. 4–5: 497–520. https://doi.org/10.1080/15387216.2016.1271345

Hojaqizi, Guliatir. 2008. "Citizenship and Ethnicity: Old *Propiska* and New Citizenship in Post-Soviet Uzbekistan." *Inner Asia* 10, no. 2: 305–22. https://doi.org/10.1163/000000008793066740

Höjdestrand, Tova. 2004. "The Soviet-Russian Production of Homelessness: Propiska, Housing, Privatisation." http://www.anthrobase.com/Txt/H/Hoejdestrand_T_01.htm.

Kalyukin, Alexander, Thomas Borén, and Andrew Byerley. 2015. "The Second Generation of Post-Socialist Change: Gorky Park and Public Space in Moscow." *Urban Geography* 36, no. 5: 674–95. https://doi.org/10.1080/02723638.2015.1020658

Kalyukin, Alexander, and Sebastian Kohl. 2019. "Continuities and Discontinuities of Russian Urban Housing: The Soviet Housing Experiment in Historical Long-Term Perspective." *Urban Studies*. Advance online publication. https://doi.org/10.1177/0042098019852326

Kornai, János. 1992. *The Socialist System: The Political Economy of Socialism.* Princeton, NJ: Princeton University Press.

Ledeneva, Alena V. 1998. *Russia's Economy of Favours: Blat, Networking and Informal Exchange.* Cambridge, UK: Cambridge University Press.

Light, Matthew. 2010. "Policing Migration in Soviet and Post-Soviet Moscow." *Post-Soviet Affairs* 26, no. 4: 275–313. https://doi.org/10.2747/1060-586x.26.4.275

Lisovskii, Vladimir G. 1983. *Leningrad: Raiony Novostroek* [Leningrad: Newbuild Districts] Leningrad, USSR: Lenizdat.

Matthews, Mervyn. 1989. *Privilege in the Soviet Union.* London, UK: George Allen and Unwin.

Morton, Henry W. 1973. "Housing." In *Handbook of Soviet Social Science Data*, edited by Ellen Mickiewicz. New York, NY: The Free Press.

– 1980. "Who Gets What, When and How? Housing in the Soviet Union." *Soviet Studies* 32, no. 2: 235–59.

– 1984. "The Contemporary Soviet City." In *The Contemporary Soviet City*, edited by Henry W. Morton and Robert C. Stuart, 3–24. London, UK: Macmillan.

Müller, Dieter K. 1994. "St Petersburg: Rysslands flaskhals på väg mot marknadsekonomi" [St Petersburg: Russia's Bottleneck on Its Way to Market Economy]. *Geografiska Notiser* 52, no. 3: 144–53.

Nove, Alec. 1986. *The Soviet Economic System*, 3rd ed. Boston, MA: Allen and Unwin.

Parkins, Maurice Frank. 1953. "Soviet Policy on Urban Housing and Housing Rent." *Land Economics* 29, no. 3: 269–79. https://doi.org/10.2307/3144834

Petrov, Nikolay, Maria Lipman, and Henry Hale 2013. "Three Dilemmas of Hybrid Regime Governance: Russia from Putin to Putin." *Post-Soviet Affairs* 30, no. 1: 1–26.

Piirainen, Timo. 1997. *Towards a New Social Order in Russia: Transforming Structures and Everyday Life.* Dartmouth, UK: Aldershot.

Pravitelstvo Sankt Peterburga. 2016. *Gosudarstvennaya programma Sankt Peterburga.Obespechenie dostupnym zhilëm i zhilishchno-kommunalnymi uslugami zhiteley Sankt Peterburga na 2015–2020 gody* [St. Petersburg's State Program: Providing Affordable Housing and Housing-Related Public Services to the Residents of St. Petersburg 2015–2020.]. http://gov.spb.ru/gov/otrasl/gilfond/gosudarstvennaya-programma-sankt-peterburga-obespechenie-dostupnym-zhi/

Reimann, Bettina. 1997. "The Transition from People's Property to Private Property." *Applied Geography* 17, no. 4: 301–14. https://doi.org/10.1016/s0143-6228(97)00023-4

Sawicki, Stanislaw. 1977. *Soviet Land and Housing Law.* New York, NY: Praeger.

Sjöberg, Örjan. 2014. "Cases onto Themselves? Theory and Research on Ex-socialist Urban Environments." *Geografie* 119, no. 4: 299–319.

Sosnovy, Timothy. 1959. "The Soviet Housing Situation Today." Soviet Studies 11, no. 1: 1–21. https://doi.org/10.1080/09668135908410184

Stan, Lavinia. 2006. "The Roof over Our Heads: Property Restitution in Romania." *Journal of Communist Studies and Transition Politics* 22, no. 2: 180–205. https://doi.org/10.1080/13523270600661011

Sýkora, Luděk, and Stefan Bouzarovski 2012. "Multiple Transformations: Conceptualising the Post-Communist Urban Transition." *Urban Studies* 49, no. 1: 43–60. https://doi.org/10.1177/0042098010397402

Szelényi, Iván. 1983. *Urban Inequalities under State Socialism.* New York, NY: Oxford University Press.

Trumbull, Nathaniel S. 2014. "Restructuring Socialist Housing Estates and Its Impact on Residents' Perceptions: 'Renovatsiia' of Khrushchevki in St. Petersburg, Russia." *GeoJournal* 79, no. 4: 495–511. https://doi.org/10.1007/s10708-014-9534-1

Ungern-Sternberg, Dorothee von, N.S. Palčikov, and R.B. Tov. 1993. "St Petersburg: Wohnungsprobleme und Wirtschaftslage in einer russischen Stadt" [St Petersburg: Housing Problems and the Economic Situation in a Russian City]. *Geographische Rundschau* 45, 3: 153–9.

Zavisca, Jane. 2012. *Housing the New Russia.* Ithaca, NY: Cornell University Press.

PART TWO

**Planning and Architecture: Designing
Socialist and Post-Socialist Urbanisms**

5 In the Spaces between Buildings: The Case of South City, Prague

STEVEN LOGAN

In 1954, Soviet President Nikita Khrushchev gave a speech extolling the virtues of the industrialization of building. Khrushchev ([1954] 1963, 161) discussed the importance of prefabrication, standardized designs, and building with reinforced concrete, claiming that "given concrete, electric motors, and lifting cranes, and other machinery – it is impossible to continue to work in ancient ways." The architects of socialist realism claimed that they were rejecting the "dull 'box style' characteristic of modern bourgeois architecture," he said, but were "decorat[ing] building facades excessively ... thus wasting state funds" (171). This turn away from socialist realism also ignited a search for new socialist models for building cities. In the 1960s, socialist city planners and architects, like their counterparts in the West, attempted to optimize free-flowing traffic, and provide rich public spaces for pedestrians, especially through elaborately designed and planned city centres. In her work on city centres in the GDR and the USSR, Elke Beyer (2011, 72) notes that the 1960s stand out as one of the last attempts to design "an all-encompassing city model for a Communist future."

The Prague housing development Jižní Město, or South City, exemplifies these two key aspects of 1960s socialist urbanism: the mass production and standardization of prefabricated apartments and the attempts to lavishly plan for efficient and high-speed mobility and separate it from a human-scale pedestrian realm. In this chapter, I show how the focus on the former often led to shortcuts in implementing the latter. In the final part of the chapter, I argue that the post-socialist transformations in South City fill in the spaces of the incomplete city of the socialist period.

Crane Urbanism

After 1964, the political atmosphere loosened in Czechoslovakia in what became known as the Prague Spring. In architecture and urbanism, increased international cooperation and a renewed optimism invigorated the search for new ways of building cities. The 1964 Prague urban plan included plans for what would be the city's three biggest postwar developments, all situated on the periphery of the city: Severní Město (North City), Jihozápadní Město (Southwest City), and South City. The main architectural journal *Architektura ČSR* no longer discussed only Soviet examples, but any new interesting projects from abroad; there was extensive coverage of Expo '67, for example. It was also easier for architects to travel abroad to see other developments. In 1966 and 1967, the chief architect of South City, Jiří Lasovský, visited the suburb of Vällingby in Stockholm, Cumbernauld in Edinburgh, and other new towns in England.[1] Many architects working in the 1960s point to the importance of the International Union of Architects (UIA) congress in Prague in 1967 (Urlich et al. 2006) that brought architects from around the world to discuss the theme "Architecture and the Human Milieu," which fit well with then-Czechoslovakian Prime Minister Alexander Dubček's idea of "socialism with a human face."

The need to construct large numbers of dwellings quickly and efficiently was still the prime concern. In a 1970 interview in the daily newspaper *Večerní Prahy*, much repeated in the press at the time, Lasovský encapsulated the task of city building: "We were to build in fifteen years what usually took 800."[2] Lasovský was referring to the medieval city of Olomouc, which at the time had a similar population to what was envisioned for South City: 80,000 inhabitants housed in 20,000 new apartments. The Pražský projektový ústav (PPÚ) was charged with overseeing the design of twenty-five of the sixty-seven new housing developments in Prague between 1957 and 1985, including South City. In a 1971 PPÚ publication, the architect and theorist Otakar Nový reiterated Lasovský's claim: "Great works, which our ancestors needed hundreds of years to make, we are today capable of managing in a few decades thanks to socialist planning, typification and industrial methods" (Nový 1971, n.p.). In the fifth (1971–5) and sixth (1976–80) five-year plans, the state planned to build over 100,000 apartments (Říha 2007, 21), while in Prague, according to a 1968 interview with Prague's chief architect Jiří Voženílek in *Večerní Praha*, the state planned to build 12,000 apartments yearly between 1971 and 1975. South City, with its 20,000 apartments, would make a significant contribution.[3] (To get a sense of the magnitude – and legacy – of these numbers, 20 per cent of the country's current housing stock was built in the 1970s.) It was up to the architects and urban planners to locate large enough parcels of land, usually on the periphery, to fulfil these demands (Hrůza 2006, 38).

The privileged actor in building cities in fifteen years instead of 800 was not the architect, but the construction crane and the state building organization. Jiří Hrůza, an influential urban planner and theorist during both the socialist and post-socialist periods, describes the practice that became known as crane urbanism (*jeřábový urbanismus*) like this: "For them [the state building organization] it was easiest if everything was made with one kind of panel, brought to the site and assembled." The crane could also not function without the rail tracks upon which it moved. According to Hrůza, "it was neither cheap nor easy to build the track for the crane, so once it was there, they wanted to assemble around it as many apartments as possible" (Hrůza 2006, 38). The placement of the crane tracks not only helped determine the arrangement of the buildings – much to the horror of the architects – but the only place where there was enough room for the tracks, and where the economies of scale and the state building quotas could be met, was on the city's periphery. Crane urbanism was a major aspect of what Katherine Lebow (2013, 4) in her work on new towns in Poland calls "state socialism's promised shortcut to modernity."

Although the drab, mono-functional socialist developments, usually referred to as the *sídliště*, came under particularly scrutiny in the post-socialist period, in the 1960s the existing developments were already being described as "grey, monotonous, dull, [and] lifeless" (Gottlieb and Todlová 1969, 211), "parasitic cities" (Nový 1971, n.p.), and "dormitory suburbs" (Hrůza 1967, 1). Hrůza (1967) describes a "crisis of the sídliště" that was not simply a crisis of the individual dwellings, but of space as a whole. In Jiří Musil's 1985 seminal study on dwelling *Lidé a sídliště* [People and the *Sídliště*], he compares life in the sídliště with life in the older, inner-city neighbourhoods of a number of different cities in then-Czechoslovakia. All aesthetic criticisms aside, people were mostly satisfied with their brand new mass-produced apartments that included many modern amenities they were not accustomed to having. And although architects and urbanists alike (e.g., Gehl 1987) have deplored the vast spaces between buildings, often a byproduct of crane urbanism, Musil's (1985, 319) respondents *appreciated* the spaces between the buildings as offering "spatial comfort" and wished they were larger. Musil also suggests that aside from the sheer quantity of housing, the postwar sídliště also introduced completely new relationships between dwelling, shopping, and open space, marked most importantly by the disappearance of the traditional city street (15). According to Hrůza (1967, 1), the common spaces of streets and squares traditionally serve both a social and a transiting function. Following a long line of modernists, he argued that developments in modern transportation meant traditional streets *were no longer an option* in planning. But in the new sídliště, where separation of pedestrian and car traffic had become the norm, all attention was given to the function of transportation – usually car, bus, and metro – while the task of replacing the social function of the traditional street had, regrettably, been forgotten.

South City as a Work of Art

In 1965, Prague's Department of the Chief Architect initiated a design competition for South City. The winning design – announced in 1967 – came from Prague architect Jan Krásný, a professor at the Technical University in Prague, and it was then passed on to Jiří Lasovský's atelier to make a detailed urban plan. In his review of the entries, Voženílek (1967) said that the site of South City was chosen carefully so it could be an urban entity unto itself, but also well-connected to the city, close to cultural and social amenities. The plans also included an area devoted to "light industry," which would provide up to 16,000 jobs, ideally for South City's residents. The development was built on the site of two already existing independent communities, Chodov and Haje, in the south-east part of Prague. In 1967, Prague widened its borders for the first time since 1921, annexing twenty-one independent municipalities, including Chodov and Haje (Borovička and Hrůza 1983, 90), and increasing its population by 70,000 inhabitants to 1.1 million people. Although Chodov and Haje became part of Prague, the planners wanted to "maximally respect" the existing communities and make an effort to connect them to the new development; however, around 200 houses would still have to be demolished to make way for South City's apartment blocks ("Hlavní architekt" 1967).

The location was also chosen so that South City could have "direct contact with the open landscape": to the west of the planned zone of light industry was the Kunratický forest park, to the south-east the Milíčovský forest, and to the north-east the Hostivař recreational area complete with an artificial lake, all within walking distance of the different neighbourhoods. The site was also chosen for its proximity to Czechoslovakia's first highway, the D1, running between Prague and Brno, which would divide South City's residential zone from the planned administrative and industry zone; work began on the highway in 1967. South City would be the gateway to Prague (Voženílek 1967, 91).

South City's conception was to a large degree governed by questions of mobility and the growing need to not only accommodate the automobile but also to control and regulate its effects spatially. The simple goal of one car for every 3.5 people set out in the official planning documentation dictated much of the look of South City. In fact, the ratio

5.1 Map of Prague (South City is now part of the Prague 11 district). Credit: Map
data from the data archive of "GIS hl. m. Prahy," licensed under CC BY-SA 4.0;
Base Map: ©IPR Praha.

exceeded the car ownership rates of any country at the time aside from
the United States and Canada, and it was not reached in Prague until
the early 1990s (Pucher 1999, 227). Although the focus on automobile
circulation as well as public transport was paramount to the visions for
South City, designers also tried to bring back the social function of the
street as a meeting place.

In "Sociological Notes on South City" in *Architektura ČSR*, Miroslav
Gottlieb and Markéta Todlová (1969) place South City's plans within the
currents of sociological thinking on the city, drawing largely on work
from the West including Lewis Mumford, Alexander Mitscherlich, and
Rene Kaes. Gottlieb and Todlová argue that the sídliště needed lively
pedestrian streets where people could sit outside at cafes and restaurants, and not just "corridors" that channel people from one destination
to the next (213).

Jiří Lasovský took up this task of providing the possibility for such
spaces. In response to the "deurbanizing tendency of the Prague
sídliště with their atomizing division of functions," Lasovský (1982,

n.p.) looked for "another philosophical concept of the city" on which to base South City's designs. Lasovský was offering an implicit critique of the Athens Charter, the modern architecture and urbanist manifesto that developed out of the 1933 meeting of the Congrès internationaux d'architecture moderne (CIAM) but was was largely authored by Le Corbusier. The charter's basis was the separation of functions: dwelling, recreation, work, circulation. This critique had been forming from within CIAM since the 1951 meeting on "The Heart of the City," which focused on the need for multifunctional, pedestrian-oriented city and suburban centres (Tyrwhitt, Sert, and Rogers 1952). The UIA congress in Prague in 1967 also criticized CIAM and its focus on separated functions as well as "minimum dwellings," the small, standardized apartments that were the subject of CIAM's 1929 meeting (Černý et al. 1967, 28–9). Lasovský's goal for South City was multifunctional centres and work close to dwellings, but also shops and cafes easily reachable on foot on what he called the *obytná ulice*, or habitable street.

The pedestrian-habitable streets, according to official planning documentation, were to be a new way of approaching the Prague sídliště, becoming "the distinct element in the formation of the environment" ("Schválení" 1968). It was on these streets that Lasovský believed the possibility for spontaneous encounters would emerge: places to play chess, places to sit and read, "places for people-watching" (Interview). The habitable streets between and around the buildings would also allow people to follow the stretches of greenery right out to the forests or the recreational area without ever having to cross a major road. In their description of the detailed urban plan, Lasovský and fellow architects Jan Krásný and Miroslav Řihošek (1969, 446) wrote that the best way to minimize the effects of car use – noise, pollution, collisions – was the "complete segregation of pedestrian and transportation spaces."

Each of South City's four neighbourhoods would have their own centre situated above the main thoroughfares, entailing a complex network of above-ground pedestrian-only walkways and shops. In an attempt to create a diverse environment, a different collective of architects would design the four neighbourhoods of South City and each would each use different building materials and technologies – brick or poured concrete, for example (Lasovský, interview). The main city centre, South City's downtown, would be located at the Opatov (formerly "Friendship") metro station. (Construction on the Prague metro extension out to South City began in 1975 and it was completed and opened in 1980.)

The city centre was to be the main element in the South City plan, worthy of CIAM's plea to architects and urban planners to attend to the heart of the city. With the city centres of Vällingby and Cumbernauld in mind, Lasovský came up with a plan for the city centre along with set designer Josef Svoboda, who played a key role in the art direction of the very popular Czech multimedia installations at Expo '67 in Montreal. Lasovský and Svoboda's plans for the city centre are well documented in a 1978 article entitled "Centre for 100,000" (Novotný 1978). This article, like those before it, contrasts the sídliště as a strict product of industrial technologies and as the vision of a new city: "Ten years ago, when the project originated ... it imagined building a city" and yet the completed areas "are still reminiscent of a sídliště" (4). The most prominent feature of the proposed centre was that it negated "all the laws of functional zoning," combining shopping centres, cinemas, cultural centres, a public square with a retractable domed roof, a factory, "a hot-dog stand next to a jewelry shop" (4), administrative buildings, and a residential area with low- and high-rise buildings entitled "Habitat," a nod to Moshie Safdie's Habitat at Expo '67. The entire centre was to be for pedestrians only, "without a single conflict with other transportation" and with a variety of pedestrian routes to choose from. South City's centre would be a "transportation paradise." The centre would affirm South City's role as mediating the relationship between the city and the countryside. Its position next to the highway would allow South City to be a "go-between" or "intermediary city" for people from surrounding villages to come and shop. With its city centre it could be a "centre of lesser importance for the suburban surroundings" (Gottlieb & Todlová 1969, 214). Once it was established that the metro would be extended to South City, its planners also envisioned it as a place for people to park their cars and then head to Prague, or a place for people coming or leaving Prague to "have a break or even spend the night."[4]

The Incomplete City?

"Only visions": this is how Vítězslava Rothbauerová describes the many ideas for South City.[5] Rothbauerová was a participant in the initial design competition and the chief architect of "South City II," a later additional residential development built in the 1980s to take full advantage of the metro. The pent-up postwar demand for housing, according to Rothbauerová, provided a significant opportunity for architects and urbanists to design entirely new cities. With a planned, socialist economy, and the state holding all the land, this process was made even easier and attractive. But the opportunity came about precisely because of the pressure to meet the state housing quotas for production. Meeting that goal required the mass production of the pre-fabricated concrete

5.2 Jiri Lasovský, South City's chief architect, proposed that this mound of excavated earth on the site of South City's Central Park would be moulded and sculpted into small hills and valleys mimicking the Czech landscape. The excavated earth, surrounded by the *paneláky*, is the lasting image of South City's crane urbanism. Credit: courtesy of Jiri Lasovský.

slabs that make the *paneláky*. Rothbauerová's own experience in the 1970s is illuminating:

> Together with the suppliers and investors, we agreed to develop a new kind of panel building for this [a second South City] development. We were excited. Of course, this promise was quickly broken as the construction workers already knew by heart how to assemble these buildings and they were not going to burden themselves by having to actually look at the drawings. (Interview)

The actual building of South City began in 1972 in the aftermath of both the Soviet invasion and occupation of Czechoslovakia in 1968 and the purges that followed. Several architects were forbidden from teaching and working, in particular Prague's Chief Architect Jiří Voženílek and Jiří Novotný, the organizer of the UIA congress. Lasovský was fired as chief architect of South City and his atelier was disbanded – they had been printing anti-occupation material – although he was still allowed to work on the city centre (Lasovský, interview).

The state construction company ceased to "coordinate on the ground" with the architects and with the PPÚ (Lasovský 1982, n.p.). Although Lasovský and the other architects had imagined different building methods and material for each of the neighbourhoods, only the concrete prefabricated panels were used, and the only difference between buildings was length – the ideal was the infinitely long building with almost no variation from one section to the next – and height: four, six, eight, or twelve storeys. Lasovský points out that the unbelievably long buildings that characterize South City were often a result of economizing on costs: it was cheaper to keep building, or to build in the immediate vicinity of the crane, than to have to move the track upon which the crane moved (Lasovský, interview).

Although Lasovský proposed a strict separation between cars and pedestrians, he envisioned within those pedestrian spaces intertwined green spaces and dwelling spaces, with greenery on the periphery stretching to the doorsteps of the buildings. But to build the tracks for the crane, entire swaths of land had to be bulldozed, frustrating the architects' plans to have people walk out of their houses and directly onto a network of pathways surrounded by greenery (Blažek 1998, 41). This is what gave so many of the sídliště their characteristic barren look in their early years – in South City, the pile of earth excavated during construction and sitting in the middle of what was to be Central Park was a material reminder of the force and speed with which crane urbanism operated (figure 5.2).

In an article criticizing South City's realization, Lasovský wrote that the hallmark of the plan, the habitable streets with preference given to pedestrians, "disappeared," and that the overall composition of the plan – that is, the contrast between the densely built-up centres and the open spaces of the periphery – was lost because of the "considerable pressure of the building technology, deforming the attempts at city building" (Lasovský 1982, n.p.). Although its designers hoped that these habitable streets would be lined with small shops and cafes, in the period of normalization there was "little street life, and so it was easier to have everything in one big building" (Rothbauerová, interview).

The attention to building housing to the exclusion of anything else save for the most essential infrastructure is one of the defining

5.3 A typical pedestrian space in South City, neglected, a space for quickly passing through. Credit: Steven Logan.

features of many of the postwar developments. The developments were intended to be self-sufficient, with not only the most necessary social services – like hospitals, clinics, and schools – but also shops, cafes, cinemas, sport facilities, and so on. However, socialist city scholar Sonia Hirt (2012, 87) notes that because of the lack of financial resources most of these latter services were not built, leading Sofia's residents to call these large dwelling complexes "un-complex complexes." This was especially the case with South City because the infrastructure costs alone were immense, in particular for building the metro out to South City. Thus, the projects had to economize "at any costs" and the victims were usually the details of the urban plans (Maier, Hexner, and Kibic 1998, 55). Hirt (2012, 40) notes that the socialist cities often suffer from a lack of small retail and "in-between" public spaces like cafes, small shops, or yards, and not just overt, grand public spaces like "friendship parks," large cultural centres, and so on.

The neighbourhood centres of South City were rarely constructed as planned, creating many uninhabitable spaces (figure 5.3), nor did the main city centre materialize, at least as Lasovský and Svoboda imagined it. In the absurd fashion that befits the era of normalization, Lasovský claims that while they were designing the city centre they knew it was never going to get built because there was only enough grant money to produce the models, but no funds to actually realize the plans. Today, the site they chose for South City's city centre remains a metro station, a few kiosks, but nothing more.

Conclusion: Where Capitalist Growth Meets the Socialist Crane

South City's very name implies incompleteness – it's a reference to an elsewhere that South City needs in order to exist. (A more accurate translation of *Jižní město* would be Southern City.) From the beginning and well into the 1970s South City was defined by both its architects and, increasingly, its inhabitants in relation to an elsewhere, the centre of Prague. The separation of traffic and the neglect of existing pedestrian spaces compounds its incomplete character: a neglected park, an unused pedestrian walkway, or any one of the local centres, which exemplify the modern practice of separating out pedestrians from vehicles. And yet Voženílek, perhaps offering an apology for the state of the socialist city, believed that cities are always unfinished; they need to grow, and so spaces should not be evaluated right at the moment they are built.[6]

Although the space of the proposed city centre remains undeveloped, the post-socialist changes in South City take advantage of the area's unfinished character to build on the project begun in the 1960s. The once barren spaces and unfinished pathways that became the dominant images of everything wrong with crane urbanism have now grown into lush green spaces (figure 5.4). The open green spaces and the "loose" arrangement of buildings that Lasovský criticized and associated with crane urbanism, appears otherwise through the eyes of residents as a space of comfort. In the online documentary *Out My Window* (Cizek, n.d.), featuring apartment buildings from around the world, South City, seen through the eyes of one family, is depicted as a successful project of post-socialist regeneration. The small shops on the streets between buildings that began to materialize in the post-socialist period as communal storage rooms for bikes and baby carriages were transformed into small shops serving the neighbourhood. These developments, however, pale in comparison with the rampant growth in shopping malls and hypermarkets in the former Eastern Bloc countries, continuing the postwar practice in the sídliště of "everything in one big building" (though this was also common to postwar suburbanization more generally).

5.4 Life between buildings in present-day South City. Credit: Steven Logan.

On the west side of the D1 highway, two significant projects have given South City a centre, although not as the planners may have envisioned. In 1984 and again in 1989, Rothbauerová prepared a study for a large-scale community centre named after the 1920s leftist avant-garde group Devětsil; however, the site was eventually chosen for one of the country's largest shopping malls, Centrum Chodov. At the same time, an international office park called The Park, comprised of mostly high-tech businesses, was being planned adjacent to the shopping mall. Both developments take advantage of their proximity to the metro.

In the late 1990s and early 2000s shopping malls, hypermarkets, big-box retail, home improvement stores, and warehousing and logistics proliferated throughout Prague, particularly in the suburbs (Sýkora and Ouředníček 2007, 219). By the mid-2000s, there was more shopping centre space per person in Prague and Warsaw than in Paris and London (Stanilov 2007, 90). Although many of these developments happened near new suburban developments or outside of the built-up areas of the city, many were also "edge-of-city" developments that nestled within the existing largely residential complexes from the socialist

5.5 Post-socialist transformations in South City – the manicured spaces of an international office park. Credit: Steven Logan.

period and within existing transit connections. Sýkora (2007) uses the term "edge-of-city" to distinguish developments in places like South City on the edge of the existing city from "leap-frog developments further out in the suburbs" (142) that lack transit connections. By 2006, construction finished on The Park, an international office park, with glass facades fronting onto groomed walkways (figure 5.5). With businesses like IBM, Dell, Sony, and GE, it has become "a symbol of the nation's twenty-first century business scene" (128). Although it does not compare in size and scope with South City's original plans for light industry, along with Centrum Chodov it establishes South City as a centre on the periphery, similar to that imagined by the architects, planners, and sociologists in the 1960s. The other major office park development of the early 2000s in Prague, Avenir Business Park, and Nové Butovice Office Park are situated in Southwest City, one of the other three major towns planned in 1964.

Although the shopping mall and office park herald the arrival of capitalism in the socialist city, they also easily situate themselves within the (unbuilt) spaces of the socialist city. In this way, the post-socialist developments in South City offer both continuation *and* rupture with socialist planning and architecture. South City is the site of competing and complementary symbols of twentieth-and twenty-first-century urbanism. Standing on the pedestrian bridge overlooking Prague's main highway connecting the shopping mall to the west to the mass-produced socialist housing to the east, one gets a particularly visceral experience of the "post" and the "socialist" in the post-socialist city.

NOTES

1 Interview with Jiří Lasovský, Prague, 24 June 2012.
2 "Beseda Večerní Prahy o Výstavbě Jižního Města: 20 000 Bytů v Zeleni," *Večerní Prahy*, 4 June 1970.
3 "Pravda o Praze," *Večerní Prahy*, 11 April 1968.
4 "Beseda Večerní Prahy o Výstavbě Jižního Města: 20 000 Bytů v Zeleni," *Večerní Prahy*, 4 June 1970.
5 Interview with Vítězslava Rothbauerová, Prague, 8 June 2012.
6 Interview with Jiří Musil, Prague, 3 July 2012.

REFERENCES

Beyer, Elke. 2011. "Planning for Mobility: Designing City Centres and New Towns in the USSR and the GDR in the 1960s." In *The Socialist Car: Automobility in the Eastern Bloc*, edited by Lewis H. Siegelbaum, 71–91. Ithaca, NY: Cornell University Press.
Blažek, Bohuslav. 1998. "Sídliště: Zracadlo nastavené době." *Umění a řemesla* 4: 41–8.
Borovička, Blahomír and Jiří Hrůza. 1983. *Praha: 1000 let stavby města*. Prague, Czech Republic: Panorama.
Černý, Milos, Miroslav Gottlieb, and Otakar Nový. 1967. *Architecture and Habitable Environment*. Prague: Czechoslovak Academy of Sciences. Published in conjunction with the exhibition of the same name, shown at the 9th UIA Congress in Prague.
Cizek, Katerina. n.d. *Out My Window: Interactive Views from the Global Highrise*. National Film Board. http://outmywindow.nfb.ca/#/outmywindow.
Gehl, Jan. 1987. *Life between Buildings: Using Public Space*. New York, NY: Van Nostrand Reinhold.
Gottlieb, Miroslav, and Markéta Todlová. 1969. "Sociologické poznámky k jižnímu městě." *Architektura ČSR* 4: 211–14.

Hirt, Sonia. 2012. *Iron Curtains: Gates, Suburbs, and Privatization of Space in the Post-Socialist City*. Hoboken, NJ: John Wiley & Sons.

Hlavní architekt města Prahy. 1967. "Zpráva o výsledku II. fáze omezené neanonymní soutěže na ideové urbanistické řešení 'Jižního města,' sídliště hl. m. Prahy." Records of the Meeting Minutes of the Bureau, Council and Local Authorities ÚNV, NVP a HMP (1945–94), Archival record, 16 May, Prague City Archives.

Hrůza, Jiří. 1967. "Krize sídliště." *Architektura ČSR* 13, no. 4: 1.

– 2006. "Jiří Hrůza." Interview by Oldřich Ševčík and Lenka Popelová. In *Šedesátá léta v architektuře očima pamětníků*, edited by Petr Urlich et al., 30–42. Prague, Czech Republic: nakladatelství ČVUT.

Khrushchev, Nikita. (1954) 1963. "On Wide-Scale Introduction of Industrial Methods, Improving the Quality of and Reducing the Cost of Construction." In *Khrushchev Speaks: Selected Speeches, Articles, and Press Conferences, 1949–1961*, edited by Thomas Whitney, 153–92. Ann Arbor: University of Michigan Press.

Krásný, Jan, Jiří Lasovský, and Miroslav Řihošek. 1969. "Podrobný užemní plán Jižního města v Praze." *Architektura ČSSR* 7/8: 442–8.

Lasovský, Jiří. 1982. "Naděje a skutečnost." *Československý architekt* 15/16: n.p.

Lebow, Katherine. 2013. *Unfinished Utopia: Nowa Huta, Stalinism, and Polish Society, 1949–56*. Ithaca, NY: Cornell University Press.

Maier, Karel, Michal Hexner, and Karel Kibic. 1998. *Urban Development of Prague: History and Present Issues*. Prague, Czech Republic: ČVUT.

Musil, Jiří. 1985. *Lidé a sídliště*. Prague, Czech Republic: Nakladatelství Svoboda.

Novotný, Jan. 1978. "Centrum pro 100,000." *Československý architekt* 15: 4–5.

Nový, Otakar. 1971. "Introduction." In Pražský projektový ústav, *Architekti Praze*. Prague, Czech Republic: HDKN Praha.

Pucher, John. 1999. "The Transformation of Urban Transport in the Czech Republic, 1988–1998." *Transport Policy* 6: 225–36. https://doi.org/10.1016/s0967-070x(99)00023-2.

Řiha, Cyril. 2007. "V čem je panelák kamarád: Kvantitativní ohledy kvalit české panelové výstavby 70. let." In *Husákova 3 + 1: Bytová kutura 70. let*, edited by Lada Hubatová-Vacková and Cyril Říha, 17–38. 2007. Prague, Czech Republic: VŠUP.

"Schválení podrobného územního plánu Jižního Města na území Chodov a Háje." 1968. Records of the Meeting Minutes of the Bureau, Council and Local Authorities ÚNV, NVP a HMP (1945–1994), Archival record 28 December, Prague City Archives.

Stanilov, Kiril. 2007. "The Restructuring of Non-Residential Uses in the Post-Socialist Metropolis." In *The Post-Socialist City: Urban Form and Space Transformations in Central and Eastern Europe after Socialism*, edited by Kiril Stanilov, 73–99. Dordrecht, The Netherlands: Springer Verlag.

Sýkora, Luděk. 2007. "Office Development and Post-Communist City Formation: The Case of Prague." In *The Post-Socialist City: Urban Form and Space Transformations in Central and Eastern Europe after Socialism,* edited by Kiril Stanilov, 117–45. Dordrecht, The Netherlands: Springer Verlag.

Sýkora, Luděk, and Martin Ouředníček. 2007. "Sprawling Post-Communist Metropolis: Commercial and Residential Suburbanization in Prague and Brno, the Czech Republic." In *Employment Deconcentration in European Metropolitan Areas: Market Forces versus Planning Regulations,* 209–33. Dordrecht, The Netherlands: Springer.

Tyrwhitt, Jaqueline, J.L. Sert, and Ernesto Rogers, eds. 1952. *CIAM 8: The Heart of the City.* London, UK: Lund Humphries.

Urlich, Petr, et al., eds. 2006. *Šedesátá léta v architektuře očima pamětníků.* Prague, Czech Republic: Nakladatelství ČVUT.

Voženílek, Jiří. 1967. "Soutěž na Jižní město v Praze." *Architektura ČSR* 2–3: 91–9.

6 Reworking the Meanings of Urbanity in Post-Socialist Phnom Penh: Toward a Commodification of Urban Centrality

GABRIEL FAUVEAUD

The last socialist era in Cambodia ended in 1989. Since the second half of the 1990s, the arrival of foreign investors and developers as well as the increasing number of local real estate projects have rapidly transformed the landscape of Phnom Penh, the capital city. According to several observers, current urban development patterns seem starkly different from the city's past. Sylvia Nam (2011), for example, describes how several historical ruptures have marked Phnom Penh's development, resulting in a tabula rasa and the decision to turn the city into a modern Asian metropolis. Although new forms of urban space production have emerged since the 1990s, Nam's argument overlooks the fact that the basic elements of Phnom Penh's territoriality remained fundamentally the same despite drastic changes of political regimes and economic systems. By looking at how this spatial structure evolved over the middle of the twentieth century, we will see that post-socialist transitions produce political, economic, social, and territorial changes evolving at different time scales. In this context, the "transition" depends on multiple temporalities are not necessarily conflated, and we have to question the relevance of considering the post-socialist city as a readable and generalizable category of analysis.

In the rare academic studies about Phnom Penh's urban development over the past fifteen years we come across two main narratives. On the one hand, the proliferation of modern buildings and large-scale projects, as well as the rehabilitation of central areas, is seen as proof of the resilience of a city that had fallen victim to an "urbicide" perpetrated by the Khmer Rouge (Shatkin 1998).[1] In particular, the city has benefited from influences and knowledge related to urban planning circulating regionally in Asia, and more specifically in Southeast Asia (Percival and Waley 2012). On the other hand, the city's rapid transformation has resulted in unequal urban development and worsened socio-economic inequality, in part due to the increasing privatization of the production of urban spaces (Fauveaud 2014b). Although these two narratives are not antithetical and may even be complementary, scholarly arguments generally fit into either one or the other. The focus on political and economic aspects of urban transformation and on land issues overlooks some important aspects of urban change. In this context, the relation between the spatial transformation of Phnom Penh and the evolution of the political and economic framework that has shaped it have not been adequately discussed.

This chapter proposes moving beyond this dichotomy by investigating urban development from a historical and territorial perspective. By focusing on past and current changes in Phnom Penh's inner city, I argue that the transition from a socialist to a post-socialist city did not drastically alter the core of the urban centrality that has structured the city's development since the nineteenth century. This continuity becomes visible if we look at the spatial permanence hidden by sudden political and economic changes and behind the rapid changes in the urban landscape. However, this centrality has been reworked by the actions of private stakeholders whose projects modify the practices and appropriations of spaces by urban dwellers. This, I argue, represents a new dynamic of territorial symbolization under the market economy, combined with the legacy of centrality. Thus, this chapter aims to contribute to a better understanding of "post-socialist urban transitions," as well as to show that both the temporalities and spatialization of socio-cultural dynamics are key issues in the transformation of formerly socialist cities.

In the first section of this chapter, we will see that scholars have paid much attention to the political, economic, and social changes involved in post-socialist urban transitions, without considering long-term processes of spatial transformations. In the second section, we will explore various facets of urban centrality in Phnom Penh, emphasizing how its socio-cultural dimensions have been structuring urban territories over the long term. In the third section, we will go back to the three different periods of socialism in Cambodia and see why Phnom Penh is a singular example of post-socialist city. Finally, in the last section we will see how the opening to the market economy, the liberalization of land commerce, and the commoditization of space since the 1990s has reworked the inherited centrality by influencing territorialization processes, as well as by imposing a value system with a distinct and dominant social, economic, and political world view.

The Post-Socialist Urban Transition: Filling the Territorial Gap

Although scholars agree that the characteristics of "socialist cities" are not homogenous (Sheppard 2000), various authors have claimed that socialist cities in quite different parts of the world (Cuba, eastern Europe, or Asia) have some characteristics in common (Andrusz, Harloe, and Szelenyi 1996). Regardless of each city's specific – local, historical, economic, political, and spatial – characteristics, scholars generally agree that the transition from a socialist to a post-socialist city allows more room for private parties to intervene in urbanization. This leads to a demarcation of private and public spaces (Scarpaci 2000), and therefore to a transformation of the relation between social and spatial orders. Indeed, the built environment is not the simple materialization of a politico-economic system: it represents at once the result, the condition, and the reflection of the social organization of societies. In many socialist states, authorities were determined to articulate the social and spatial orders, especially in urban areas: the city had to incarnate a "societal project." Impressive public buildings and spaces had to show the magnificence of the state and illustrate the strength of the population's work force; vast public spaces, esplanades, and monuments are used for the *mise en scène* of state power and to show the "achievements" of the social revolution. Therefore, one can expect that the transition from a socialist to a post-socialist regime implies important transformations of urban spaces. Moreover, the transitional process ultimately leads to an important redefinition of socio-cultural meanings that the production of urban spaces is supposed to express.

The effects on cities of politico-economic transitions from socialist to post-socialist states have been the subject of an important body of work, especially since the fall of the Soviet Union at the end of the 1980s. However, as Geoffray (2013) has emphasized, the first studies on socialist cities certainly proposed the finest interpretation. Referring to Fischer (1962), Carter (1979), French and Hamilton (1979), and Andrusz, (1984), she points out how urban planning strategies under socialist regimes tried to standardize urban structures and link the urban and social orders (Geoffray 2013, 157). This remark is one of many examples showing that social scientists have always been more interested in the role of social, political, and economic transitions in the reorganization of the way post-socialist cities are produced rather than in the evolution of cities' spatial issues, especially prior to the spatial turn of the 1980s. K. Stanilov (2007, 3) explained this situation by arguing that "issues related to political struggles, economic development, markets, social inequalities, and class polarization have dominated the public discourse and scientific research alike."

In terms of spatial transformations, scholars have identified key processes of post-socialist transition, such as urban sprawl, multimodal urban structures, or the "revalorization" of the inner city. Downtowns have often been described as key political and social spaces of post-socialist urban transition (Sýkora, 1999). By using the terms "spatial adjustment" and "spatial realignments," Stanilov (2007, 6) sees the effects of the market economy and of the privatization of the urban spaces production as a "frivolous application of patterns of development 'borrowed' from the West" (8). The convergence of urban patterns in the world towards the Western urban experience is a classic argument of scholars studying post-socialist transitions (Tsenkova 2006), or those interpreting the changing trajectories of Global South cities since the 1980s. For instance, the spread of gated communities and Central Business Districts (CBDs) in formerly socialist cities has been interpreted as a major example of such urban logics, as well as the rise of individualistic behaviours (Blinnikov et al. 2006; Fauveaud 2016; Hirt and Petrović 2011; Hirt, 2012). In the Southeast Asian context, Rimmer and Dick (2009) have argued that urban patterns are showing signs of convergence towards Western models, with the rise of peri-urban landscapes, enclosed residential estates, malls, CBDs, polycentric models, urban corridors, and so on.

However, as Roy (2015) recently argued, if urbanization became one of the most global process of territorial changes during the last fifty years, it does not have to be interpreted as the emergence of universalistic forms of urban patterns and ways of life. On the one hand, this misunderstanding is partly due to the domination of Western theoretical frameworks formulated through the study of Western cities. On the other, it can be seen as a consequence of the global spread of neoliberalism and of the urban forms and processes that it participates in shaping, which favour the diffusion of arguments defending the idea of a global homogenization of the production processes of the city (Parnell and Robinson 2012). As a consequence, since the 1990s, scholars analysing spatial changes generated by post-socialist urban transitions have been strongly influenced by the concepts used to describe post-industrial urban transitions, such as gentrification, enclosure, or beautification, without necessarily interrogating their adequacy to local conditions of urban change. I do not argue that such theories are inadequate. However, I do argue that their non-contextualized use might hide the importance of the *longue durée* in the territorialization and materialization of post-socialist urban transitions, resulting in a confusion between results and processes.

This difference between global or universal processes is thus crucial in the context of interpreting post-socialist urban transitions. The general framework of analysis for these transitions has traditionally been drawn from experiences in Eastern Europe, although the Asian transitions have increasingly caught the attention of social scientists over the last decade. In Asia, studies of the transition from a socialist to a post-socialist city generally call attention to changes in state intervention in urban planning. Whether in Vietnam, China, or Indonesia, the urban models of "universal progress" and "social equality" have been reworked due to new articulations between the nation state and private actors within the broader framework of a privatization of urban space (Lin 2004; Hogan et al. 2012; Peyvel 2013). Scholars have also emphasized how the transition from a socialist to a post-socialist city goes hand in hand with "modernity." Since the era of "socialist modernism," a vision of urban modernity has emerged in relation to the globalization of urban and architectural forms (Diener and Hagen 2013). In many cases, scholars seem to be particularly attentive to the haste of post-socialist transitions; especially in Asia (and in China in particular), they generally insist on the fast, drastic, and sometimes dramatic reshaping of post-socialist cities (Gaubatz 1999).

The temporalities of post-socialist transitions are certainly a central issue here. The speed of spatial transformations may hide the juxtaposition of different and interrelated temporalities (social, spatial, political, etc.) that do not evolve at the same rhythm. Besides, transitional contexts always produce hybrid situations, and post-socialist transitions are made of mixtures of old and new, socialism and capitalism, public and private, and so on. Understanding the effects of politico-economic transition on urban spaces is thus a difficult task. In this chapter, we see that post-socialist transitions involve different and overlapping temporalities of urban space production: the long-term of spatial and territorial fabrics, with the short-term of the transition to market economy, capitalist modes of production of urban spaces, and the neoliberal city. Based on the assumption that the urban centrality in Phnom Penh (which I will define) represents one of the foundations of the appropriation of urban spaces by residents (what I call processes of territorialization), I will argue that spatial transformations do not necessarily led to a tabula rasa of previous territorial dynamics, but rather to a reorientation and reformulation of vernacular spatial practices and representations.

Phnom Penh: The Multiple Dimensions of the Urban Centrality

Phnom Penh is a small capital city covering approximately 670 square kilometres and counting about 2 million inhabitants. According to projections based on the last population census (conducted in 2008), urban dwellers represent about 25 per cent of the country's population, and 80 per cent of them live in Phnom Penh. In this city, rapid urbanization has led to a general reorganization of urban territories. Since the early 1990s, the installation of garment factories on the city's periphery has accelerated peri-urbanization. Starting during the second half of the 1990s, medium- and large-scale residential estates have sprung up on the periphery, and most of their new residents have moved from the inner city. In the historical centre, condominiums, skyscrapers, and modern offices proliferate. This transformation is generally understood as the result of Cambodia's rapid economic growth since the 1990s. Phnom Penh's urban dynamics have followed the main trends of urban development in Southeast Asia (Goldblum and Franck 2007): tertiary activities and corporate headquarters concentrated in the city centre (inside an emerging CBD), and new residential spaces in peri-urban areas. However, a closer look at this new relation between the city centre and its periphery shows that current urban developments do not totally break with historical urbanization trends.

Phnom Penh's development has always been structured by its location on the western bank of the natural site called the "Four Arms," which refers to the X formed by the convergence of the Mekong and the Tonle Sap rivers and then, farther downstream, by the divergence of the Mekong and Tonle Bassac rivers (see figure 6.1). This aquatic space has always represented the traditional centrality of the city. The notion of centrality evokes various processes, depending on what scholars wish to underline. In geography, centrality has traditionally referred to the dispersion and agglomeration dynamics of economic activities in cities depending on networks, locations, and situations (see Bird 1977; Castells 2010). However, centrality is also defined by various aspects of the social and cultural dimensions of spatial dynamics (Lussault 2007). In Phnom Penh, this centrality has shaped urban development through the ages; it is both physical and symbolic as well as economic and political.

The location of Phnom Penh can be attributed to the Four Arms, since this river crossroads favoured the development of trade and the settlement of merchants – it has always shaped urban development and represented a central component of the city's identity. In the Royal Chronicles of Cambodia, as well as in the journals written by explorers between the fifteenth and the nineteenth centuries, Phnom Penh is usually called Chaktomuk (literally Four Faces, another name for the Four Arms site; Phoeun 1991). The legend about the foundation of Phnom Penh in the fifteenth century places the Four Arms on a spiritual dimension. Phnom Penh literally means the "mountain or mound of Penh." According to the legend, a woman named Penh found firewood in the middle of Four Arms. While cutting it, she came upon five small statues, and decided to put them in a temple (Wat Phnom) to be built at the top of a man-made mound (Osborne 2008). Wat Phnom is an important

6.1 The Four Arms in Phnom Penh. Source: Gabriel Fauveaud, 2015.

tourist and spiritual place in the capital city, and a significant place in Phnom Penh's history.[2]

The relation between the city, Cambodian royalty, and the Four Arms was sealed during the second part of the 1800s, when Phnom Penh became the capital of modern Cambodia. When King Norodom I moved to Phnom Penh (between 1863 and 1866) after having signed the protectorate treaty with France, he built his palace on the axis of the Four Arms. Since the capital in Cambodia is historically the king's dwelling place (Lamant 1991), this installation in front of the Four Arms "fixed" the symbolic and political relations between the monarchy and the city.

The particular relationship that Cambodians have with the riverscape in Phnom Penh is expressed on several occasions throughout the calendar year (Pierdet 2005), such as during the traditional water festival (Leclèrc 1904) held every year in November to celebrate the end of the rainy season and the reversal of the Tonle Sap's course, after the river has flooded the vast agricultural plain in the centre of Cambodia.[3] The king oversaw festivities from the top of the royal pier, just opposite the Four Arms, where pirogue teams from throughout the country competed.

The extraordinary waterscape has always impressed foreigners and visitors, who during the colonial era called Phnom Penh "the pearl of southeast Asia." Today, the quays in the historic centre bordering the Four Arms form the capital's touristic heart. Young people and families like to "hang out" in cars or on motorbikes along the main roads near the river. Until recently, the riverbanks were crowded at the end of the day and during weekends as inhabitants came to picnic, take a walk, or engage in other activities (sports, dance, etc.). In recent years, however, these practices have evolved due to changes in urban policies and a global evolution of spatial practices by urban dwellers (see the last section). Despite these changes, places surrounding Four Arms are still very popular. Activities that were traditionally pursued in the quays spread in other areas facing the Four Arms, such as at the end of Chruy Changvar Peninsula or on Diamond Island (see figure 6.2). City-dwellers thus continue to appropriate the Four Arms and surrounding areas, which continues to make them the driving force of the urban centrality.

Urban centrality in Phnom Penh has largely been shaped by the Four Arms, the Royal Palace, and the historical, spiritual site of Wat Phnom. These three sites correspond to the main structures of the historical centre, and represent a vernacular "cultural landscape" (Czepczyński, 2008) that has survived and endured despite changes in urban policies and urbanization. However, the symbolization of this centrality, to rephrase Monnet (2000), evolved through time according to successive changes in the politico-economic system.

Three "Cambodian Socialisms" and Their Role in Transforming Phnom Penh

The case of Phnom Penh has to be considered as a singular example of post-socialist transition, for two main reasons. Firstly, interventions in urban development driven by post-independence socialist ideas were not as straightforward as in other socialist cities. Secondly, as we will see, the Khmer Rouge "urbicide" represented an extremist "socialist politic." Both of these reasons make the case of Phnom Penh unique, yet help shed important light on the nature of post-socialist transitions in Asia. Contemporary studies of Phnom Penh's development draw attention to four distinct periods: the colonial era (1863–1953), the socialist-modernist period (1953–70), the civil war and "urbicide" (1970–89), and the contemporary or post-socialist period (1989–today). This artificial division into historical periods errs by reducing Phnom

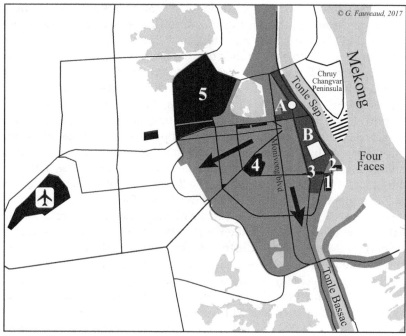

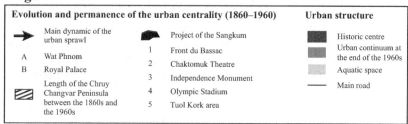

6.2 Phnom Penh's transformation under Sihanouk's "socialist" government.
© G. Fauveaud, 2017.

Penh's long history (in particular, the evolution of its spatial structure) to the transition from a socialist to a post-socialist city.

Two years after Cambodia's independence in 1953, the Sangkum Reastr Niyum (usually translated as "People's Socialist Community") won the elections. Norodom Sihanouk, the king and self-proclaimed "father of independence," had created this party, which he headed from 1955 to 1970. He advocated a "royalist-Buddhist socialism" (Vickery 1999), though the socialist nature of his politics remains a subject of contention among historians (Girling 1971). Sihanouk's socialism drew inspiration from several sources: Sukarno (Indonesia's first president), Nehru (India's first prime minister), and Zhou Enlai (the first prime minister of the People's Republic of China). He also admired de Gaulle (France), Kim Il Sung (North Korea), Tito (Yugoslavia), and Ceauşescu (Romania; Chandler 1991, 87–91). Sihanouk did not establish collectivism; property remained private. This authoritarian prince did not share political power, and he cracked down on many opponents (Osborne 1973). He also forged a special relationship with his people through speeches and the media, and by staging his accomplishments. He publicly represented himself in ways designed to gain popularity and be seen as the "father of the nation" (Abdoul-Carime 1995). The three pillars of the new Kingdom of Cambodia were race, religion, and the throne. One of Sihanouk's main objectives was to assert a new Khmer identity after French rule. At the national level, he favoured the construction of roads, railways, and a seaport, and developed industry. He fostered the educational system, and the literacy rate increased significantly during the 1950s and 1960s. He managed to obtain money and technical expertise from both East and West (in particular, France, China, the United States, and the Soviet Union). Sihanouk tried to keep Cambodia neutral in conflicts between the major international powers. But this position was ever harder to hold given, in particular, the second Vietnamese war. The rupture of political relations with the United States in 1965 preceded Sihanouk's fall in 1970, when General Lon Nol, supported by the United States, seized power.

As in many countries following independence, the quest for a new identity and for new models of economic, political, and social development led to state interventionism in the capital city. Phnom Penh was to be a socialist incarnation of independence and the manifestation of a cultural revival. As in other newly independent Asian countries, the independence movement, represented in Cambodia by the Sangkum Reastr Niyum, was influenced by modernism in architecture. Prince Sihanouk asked Vann Molyvann, a Cambodian architect educated in France who had worked with Le Corbusier, to design urban projects for the capital. International architectural contests and personal invitations brought to the city major figures of the Modern Movement, such as Vladimir Bodianski, Gerald Hanning, and Robert Hansberger, all of whom had worked closely with Le Corbusier. Between the end of the 1950s and the start of the 1960s, a new architectural movement was born, what was later to be known as "New Khmer Architecture" (Reyum 2001; Molyvann 2003; Grant Ross and Collins 2006).

The projects initiated by Molyvann and his colleagues, with support from the government, transformed Phnom Penh. Three areas, in

particular, were affected: the historical centre of the city and its immediate periphery; the west (urban sprawl); and the northwest (the lake
zone; see figure 6.2). The goals pursued in these three areas differed
but were always in line with the party's social, cultural, and political
objectives. In addition to these three major projects, several universities, institutions, and housing estates were also built under Sihanouk.

In the city centre, two main projects opened new axes, north-south
and east-west. Sihanouk ordered an "independence monument" to be
erected at the centre of a large circular square. Wide boulevards led to
the monument, and public gardens were laid out. A new perspective
linked the Four Arms, Wat Phnom, and the Independence Monument.
Sihanouk also ordered the construction of a big housing project (Front
du Bassac) along the river, on the axis running from the Four Arms to the
Independence Monument. This project, whose design was inspired by Le
Corbusier's work (e.g., La Cité Radieuse), included housing, a theatre, the
Museum of Independence, and common spaces for residents. Front du
Bassac was built to accommodate the athletes taking part in the GANEFO
games held in Phnom Penh in 1966, but after the event it hosted govern-
ment employees under a home-ownership program. Advertisements for
the project announced a "new urban centre" for the capital.

In the west, Sihanouk had a stadium built on the racecourse dating
from the colonial era. It would be used for important political events,
both national and international. As the stadium was being built, the
city expanded westwards. Urbanization in the west entailed laying out
infrastructure (roads) and public spaces (markets). In the northwest,
Sihanouk made plans for a garden city that drew inspiration from
Ebenezer Howard's projects. Along with Front du Bassac in the southeast, this new urban area was to accommodate an exponentially growing urban population.

Although Sihanouk's urban policy was intended to adapt the city's
shape to its growth and create a new centrality in the west, the transformation of Phnom Penh during the 1950s and 1960s did not fully break
with the colonial past. The New Khmer Architecture was a symbiosis of the Modern Movement and Angkorian architecture of the clas-
sical period (twelfth–thirteenth centuries). For Sihanouk, references
to Angkor's glorious past underscored the greatness of Cambodian
civilization – a process initiated by colonial officials. The French, who
claimed to have rediscovered Angkor in the nineteenth century, used
the reference to Angkorian civilization to define a "Khmer identity" at
the turn of twentieth century, when French policy replaced "assimila-
tion" with "association." As a consequence, Angkor became the symbol of what Wright (1987, 1991) has called "tradition in the service

of modernity." References to Angkor were reworked in the new cultural and political context following independence, but architectural influences and cultural meanings did not signal a full break with the colonial era (Groslier 1985; Edwards 2007; Dagens 2008). Changes in Phnom Penh following independence were continuous with the previous period's territorial dynamics. The expansion of the city westwards and southwards did not alter previous urbanization trends, and the Four Arms still represented the eastern limit of urbanization.

The Sangkum Reastr Niyum's urban planning reshaped part of the city's cultural landscape. The government, architects, and planners relied on the previous territorial organization to impel a new dynamic of urban modernization and cultural renewal. The decision to develop the southeast instead of the north represented a reappropriation of the city's historical centre and foundation. The urban projects launched by Sihanouk and the new link made between Wat Phnom, the Independence Monument, and Front du Bassac moved the city's symbolic centre southwards. This movement followed the displacement of the centre of the Four Arms as alluvial deposits made Chruy Changvar Peninsula longer (Reyum 2001). One "socialist" characteristic of Sihanouk's urban planning stands out: the decision to reclaim the Four Arms for the people (and the Sangkum Reastr Niyum), which would ultimately redefine the historical and traditional centrality of the city.

This urban policy came to a sudden halt. In 1970, Sihanouk, while on a trip to France, was ousted by General Lon Nol with the support of the United States government, which criticized the prince's closeness to communist governments, in particular China. Between 1970 and 1975, Lon Nol tried to slow down the communist expansion in the countryside and fought several Khmer Rouge battalions. On 16 April 1975, the Khmer Rouge occupied Phnom Penh and took power. During the next three days, the entire city was emptied, and all its inhabitants were forced onto the road or sent to labour camps in rural areas. Under Khmer Rouge rule from 1975 to 1979, all Cambodian cities remained uninhabited by civilians, and were only occupied by few military troops. For the new rulers, cities were places of "perversion," since many urban dwellers had succumbed to the capitalism diffused by Western powers. These "perverted" people, named by the Khmer Rouge the "New People," had to be re-educated or killed. According to the Khmer Rouge, the city was, especially at the time of its takeover, a dangerous place where outside forces and agents of the West were assembled to overthrow the revolution. For all of these reasons, they decided to annihilate it.

Nonetheless, the city's place in the new regime's agrarian revolution was ambiguous in many respects. Phnom Penh was still the capital, and remained an important political, military, and diplomatic centre under the Khmer Rouge (Carrier 2010; Tyner et al. 2014), who used the national stadium for big political ceremonies – evidence that certain urban spaces retained the same function as prior to the attempted urban annihilation.

This ambivalence towards cities did not, however, keep this "urbicide" from turning into a tragedy. When the Vietnamese army defeated the Khmer Rouge and entered the city in 1979, they discovered a ghost town, part of it destroyed. During the four years of Khmer Rouge rule, over 25 per cent of Cambodia's population died – including half of the population from the capital (Sliwinski 1995). A decade passed before Phnom Penh's population grew again to the same size it was prior to the Khmer Rouge. Owing to this anti-urban policy and the quasi-systematic assassination of intellectuals, civil servants, and skilled workers, much "urban knowledge" was lost between 1975 and 1979. The reconstitution of information and skills in matters related to urban planning and management has been a major issue since the 1980s, along with the reconstruction of infrastructure. Although it was extreme, violent, and "total," this attempt to kill a city did not fully erase previous socio-spatial and socio-cultural patterns. The key elements of urban centrality from before the Khmer Rouge continued as determinants of urban (re)development.

Between 1979 and 1989 the Vietnamese Army occupied Cambodia and the country was governed by a hybrid Vietnamese and Khmer administration. Most members of this elite had been educated in Vietnam, but a few had been trained in "brother" countries. A collectivist system was set up and private property was forbidden (even though the administration often turned a blind eye to informal business in real estate). Controls over unofficial private transactions in real estate were loosened as of 1986, in line with the Vietnamese Đổi mới reforms.

The reconstruction and repopulation of Phnom Penh during the decade following 1979 suggests, to a degree, the territorial dynamics that had prevailed prior to the urbicide perpetrated by the Khmer Rouge. In Carrier's (2007) study of changes in land rights and uses between 1979 and 1989, we notice that "places of power" from before 1975 were maintained. The historical centre remained the country's administrative centre, where decisions were made, and government buildings were re-occupied by the new administration. Most areas in the city still have the same function (residential, commercial, administrative) as before the Khmer Rouge regime. Officials in the Khmer-Vietnamese administration occupied the residential areas where officials and tycoons used to dwell before the takeover by the Khmer Rouge. However, other vacant residential areas were occupied by new social groups, as in the old Front du Bassac project. The new inhabitants were no longer state workers, but rather poor families from the countryside (Simone 2008).[4]

The effect of the post-socialist transition in Phnom Penh after the departure of the Vietnamese in 1989 and the acceleration of economic liberalization at the beginning of the 1990s are quite similar to other post-socialist cities: suburbanization, rise of private real estate projects, revitalization and beautification of central areas, eviction of poor communities from the city centre, and so on. However, these drastic, often violent, changes have not completely destroyed the inherited spatial organization of Phnom Penh, particularly some of its socio-cultural dimensions. Urban spaces surrounding Four Arms continue to play, as during the post-independence period, a key role in redefining the socio-cultural meaning of Phnom Penh's urban centrality.

Just before they left in 1989, the Vietnamese abolished the land tenure rights dating from before 1975 and reintroduced private property. Their departure signalled the end of the three successive socialist periods. Above all, it represented profound changes for Phnom Penh as Cambodia rapidly adapted to the market economy.

Redefining Centrality in the Era of the Market Economy

Since the beginning of the 1990s, the traditional centrality around the Four Arms, the Royal Palace, and Wat Phnom has been maintained but has also evolved due to the rapid opening to the market economy in the 1990s and the commoditization of land through the formalization of the private property system. Rather than erasing the previous territorial organization and its socio-cultural dimensions, the market economy added a new layer of centrality characterized by spatial reappropriations and socio-spatial exclusions.

At the beginning of the 1990s, Phnom Penh was still being reconstructed. The Paris agreement signed in 1991 between the United Nations and the Cambodian political parties provided for a peacekeeping mission from 1991 to 1993 to organize free elections. More than 20,000 foreigners, both soldiers and civilians, arrived in Phnom Penh – an acceleration of the trend set off by the arrival of international NGOs at the end of the 1980s. The arrival of foreign workers, the privatization of land ownership, and the real estate market reinforced urban centrality, notably because the city centre was where international institutions set up offices for their staff (Fauveaud 2014a).

The new constitution in 1993 provided for a market economy. The private sector controlled nearly all agricultural, manufacturing, and tertiary activities (Népote and de Vienne 1993; Gerles 2008). For the Cambodian government, in cooperation with the international community (in particular the World Bank and Asian Development Bank), one priority was to integrate the country into regional and international economic organizations. Cambodia joined the Association of Southeast Asian Nations (ASEAN) in 1999 and the World Trade Organization (WTO) in 2004 (the first country from the "less developed" group to join the WTO). It also signed eleven bilateral agreements on investment with Southeast Asian countries in the two decades after 1990, and became a member of subregional organizations, such as the Greater Mekong Subregion and the ASEAN Mekong Basin Development Cooperation (AMBCD). Membership in these organizations depended on liberalizing the country's economy.

Several laws passed in 1994 and 1995 laid the groundwork for local and international investments, which have created a macroeconomic environment favourable to private enterprise. In 1992, a new land law defined types of property. In 1994, a law on urban planning and management was adopted and in 1997 another law introduced building permits. The Ministry of Land Management, Urban Planning, and Construction was set up in 1999. In the two years following the passage of a new land law in 2001, several decrees were issued that set the conditions for various types of property, land uses, and land management, as well as for real estate businesses, both public and private. Access to land ownership was a major issue in this process.

Phnom Penh remained the centre of the nation's economy, and considerable foreign investments since the 1990s have helped transform the city. Starting in the mid-1990s, garment factories, mostly owned by Asian companies, were set up on the city's periphery. More than 350 factories now employ more than 350,000 workers, most of them coming from the countryside and living near the factories where they work. Local and international institutions and NGOs have set up their headquarters in the capital, further boosting the real estate market. A substantial part of the flow of foreign capital through aid to development (De Vienne 2012) or foreign investments is redirected towards real estate, though it's difficult to quantify. From the late 1990s until 2010, land prices shot up and speculation fuelled urbanization in peri-urban areas where new housing estates have sprung up, and several large-scale urban projects have been launched (see figure 6.3). All of this has impelled urbanization and transformed the relationship between the city's centre and peripheral areas (Fauveaud 2015).

Two new business districts have sprung up, one in the north of the historical centre (the west part of the colonial sector), and one opposite the Four Arms. The one in the north represents continuity with a trend inherited from the colonial era. This area has traditionally been devoted to tertiary activities (banks, luxury hotels, embassies, and big institutions

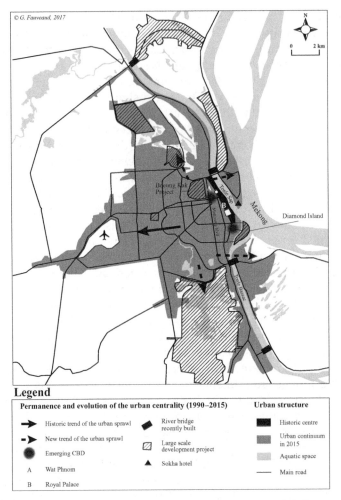

6.3 The contemporary evolution of Phnom Penh's urban centrality. © G. Fauveaud, 2017.

both local and international). This inheritance was reinforced at the end of the first decade of the new century, when two major government buildings (the Prime Minister's Office and the Council of Ministers) were built there as well as the city's first (two) skyscrapers. The Boeung Kak project further confirms this trend (see figure 6.3). Not only does this business district profit from the gardens and boulevards laid out during the colonial era, it also lies close to the riverbank, which, since colonial times, has been an axis of leisure and business (not to forget its function as a landscape). It also has one of the city's main boulevards, Monivong.

Second, a subcentre created in the southeast continues from the new centre there that dates from the period following independence, when Front du Bassac was built. The civil war and the Khmer Rouge put a halt to the ambitions of Sangkum architects and urban planners. Part of the Front du Bassac was occupied by poor families beginning in the 1980s and part of the land was sold to private investors in the 1990s. One of the buildings there was bought in 2008 by a Malaysian company to use for offices. The other was sold in 2016 to a Japanese company, and the inhabitants were evicted with compensations in 2017. The emblem of this development in the southeast is Diamond Island (see figure 6.3), currently being built by the Overseas Cambodia Investment Corporation (OCIC), which acquired the land in 2006 on a long-term lease and became the owner in 2009. This project is on an alluvial island that emerged from the river during the 1970s. The OCIC has levelled the new island and enlarged it. The project includes a residential zone (Elite Town), a convention centre, and areas for business and leisure. Several other residential, commercial, and office projects have been designed for Diamond Island since then. This subcentre benefits from its proximity to major government buildings (e.g., the National Assembly, Senate, and Ministry of Interior) and embassies. This has spurred the construction of gated communities for government employees, embassy personnel, politicians, and tycoons. This new subcentre has been reinforced by the construction of other office buildings and commercial zones (in particular the Aeon Mall).

The multiplication of private urban projects in the southeast is a factor that pushes for a redefinition of the symbolic meaning of centrality dating back to the nineteenth century, when the king settled in Phnom Penh. As we have seen, Sihanouk's "socialist" urbanism could be perceived as an attempt to redefine the Cambodian people's place in the symbolic relation between the city and the Four Arms. In contrast, the relation between the city and the "aquatic centrality" in the southeast is being redefined in a post-socialist context by urban projects conducted by the private developers who own land around the Four Arms. Seen in this way, Diamond Island is a key to redefining symbolic centrality in Phnom Penh. The many private urban projects being conducted around this site reinforce this process. The recent construction of a luxury hotel at the far end of Chruy Changvar Peninsula (Sokha Hotel, see figure 6.3) is one of the best examples. Abetted by real estate speculation on the eastern bank of the Four Arms, this urban process is sure to continue during the coming decades.

As a consequence, and given the other residential and business projects in the area, the relation between the city and the Four Arms is defined in today by the actions of private developers for whom this natural site's incredible landscape represents an added value for investments. Developers try to attract buyers by playing on the symbolic relation between the city's residents and this "aquatic centrality"; one advertisement for Diamond Island emphasizes the "water culture and beyond."

The transition to a market economy in Cambodia and the commoditization of space that accompanied it thus rely on a renewed symbolization of vernacular territorial representations and practices. Because the other key elements of urban centrality remain untouched (the royal palace and the spirituality of Wat Phnom), the historic urban centrality is not being denied but reformulated, thus signalling both continuity and change in urban development, beyond political and historical ruptures. This shows the multi-temporal dimensions of urban transformation, and the fact that spatial dynamics may endure more than social, political, and economic environments.

However, this symbolization process is far from being neutral: it is also the result of a materialization of dominant social (upper social groups), economic (capitalism and liberalism), and political (in Cambodia, neo-patrimonialism and cronyism) values and representations. Besides, this process has been reinforced by local authorities who are trying to normalize socio-spatial practices of inner-city public spaces. The municipality has recently passed a decree that limits, and in some areas forbids, the occupation of open green spaces in the city centre. Officials declared that these areas, because of their high landscape quality, have to be preserved from the damage caused by urban dwellers who are accused, for instance, of dropping their waste and damaging the grass. Unofficial and "off-the-record" discourses are quite different. The occupation of these spaces for picnics or for leisure activities (ball games, etc.) do not correspond to the modern urban landscape that officials want to promote. For some, these popular appropriations of public spaces evoke rurality more than urbanity, and have to be regulated or proscribed. However, it would be wrong to interpret this decision as a simple desire by local authorities to exert unilateral repression on lower-status social groups. The discourses and attitudes of the municipality correspond to a broader representation shared by many Phnom Penh inhabitants: central neighbourhoods in the touristic part of the city and the quays have a bad reputation: they draw bars, drunks, prostitutes, beggars, and foreigners with bad – or just strange – behaviour. As a consequence, the riverbanks are less attractive than before. This recent evolution of the relation between urban dwellers and urban centrality also explains the success of Diamond Island as a "public space" in a private project, where Cambodian families and youth have almost complete liberty to appropriate these urban areas as they want.

More generally, however, the redevelopment of the central area has resulted in an exclusion of the poor because of the rising price of real estate and the official policy of the centre's "beautification" (Springer 2010). Land seizures, socio-economic exclusion, and political repression are all part of a broader metropolitan strategy that serves the interests of important tycoons, officials, and clans who benefit from maintaining the social order (Fauveaud, 2015). In order to produce a modern, beautiful, safe, and clean central tourist area, the central government (especially the Ministry of Social Affairs and the Ministry of Interior) regularly arrests beggars, homeless people, and prostitutes and deports some of them to "social centres" – some resembling real prisons – on the periphery, where they are forcibly detained for a few days or weeks. This oppressive system attests to the authoritarian nature of Phnom Penh's modernization and metropolization process (Fauveaud, 2014b; Springer, 2016).

Conclusion

In this chapter, I showed that the transition from socialist to post-socialist cities is made of multiple time scales. Using the example of Phnom Penh, I have argued that drastic and sudden political and economic changes in post-socialist contexts do not automatically break with pre-existing territorial dynamics. With regards to the existing literature on post-socialist cities, I stressed the importance of paying attention to spatial changes and reterritorialization processes, rather than focusing exclusively on the economic, social, and political aspects of post-socialist transitions. To do so, I chose to focus on the transformation of Phnom Penh's inner city: the central areas representing strategic spaces of urban transition (post-socialist, post-industrial, etc.). I used the notion of "spatial symbolization" to demonstrate that the transition to a market economy represent the addition of another layer of territorial changes, and a reformulation of former spatial dynamics. The case of Phnom Penh is particularly useful because the city was the object of an attempted "urbicide" by the Khmer Rouge regime.

Thus, the current urban development patterns in Phnom Penh, certainly new in terms of their forms, must also be studied with regards to the *longue durée* of spatial changes inherited from the past. This continuity can be understood in the permanence of the key elements of Phnom Penh's urban centrality: the river site, royalty, and spirituality. These three dimensions of urban centrality correspond to the linkage between three places: the Fours Arms, the royal palace, and Wat Phnom. Political changes, though tragic, have not drastically altered the fundamental spatial structure that has ultimately survived.

Having said that, aspects of the cultural landscape and its meanings, as well as spatial practices and representations evolved due to the commoditization of space, the multiplication of private real estate projects, and the enforcement of new coercive laws. In this sense, the symbolization process of the market economy in the inner city is also a projection of dominant social values, and of the will of those who have the power to transform the space (private developers, municipal authorities, upper socio-economic groups, etc.). Land seizures, socio-economic exclusion, and political repression have ignited social protests in Phnom Penh since 2010. Public parks in the historical centre are strategic places for sit-ins and demonstrations. Many of these public protests end in front of symbolic places of power in the city centre, such as the prime minister's house or the National Assembly. The inner city has become a space of political and social confrontation, which will certainly participate in redefining other aspects of the urban centrality in the near future.

NOTES

1 This Maoist organization emerged as a political force in the countryside during the 1950s and gained support until it took power in 1975 and ruled until the intervention of the Vietnamese army in 1979.
2 The temple contains the ashes of Ponhea Yat, who, according to the Royal Chronicles, was the first king to establish the capital in Phnom Penh, in the fifteenth century.
3 The course of the Tonle Sap changes twice a year. During the rainy season from May to November, because of precipitation from the annual monsoon and melting snow in the Himalayas, its course runs from the Mekong to Tonle Sap Lake. During the dry season from November to May, it runs from the lake to the Mekong. Four Arms represents the heart of this peculiar hydrological system, which largely sustains Cambodian agriculture.
4 The last families living in the Front du Bassac were evicted in 2017 after the land was sold to a Japanese company.

REFERENCES

Abdoul-Carime, Nasir. 1995. "Réflexion sur le régime Sihanoukien. La monopolisation du verbe par le pouvoir royal." *Péninsule* 31, no. 2: 77–97.

Andrusz, Gregory. 1984. *Housing and Urban Development in the USSR*. London, UK: Macmillan.

Andrusz, Gregory, Michael Harloe, and Ivan Szelenyi, eds. 1996. *Cities after Socialism: Urban and Regional Change and Conflict in Post-Socialist Societies*. Oxford, UK: Blackwell.

Bird, James. 1977. *Centrality and Cities*. London, UK: Routledge.

Blinnikov, Mikhail, Andrey Shannin, Nikolay Sobolev, and Lyudmila Volkova. 2006. "Gated Communities in the Moscow Greenbelt: Newly Segregated Landscapes and the Suburban Russian Environment." *GeoJournal* 66: 65–81. https://doi.org/10.1007/s10708-006-9017-0

Carrier, Adeline. 2007. "Les 'lois de la possession' à Phnom Penh: Conversion des droits d'usage résidentiel issus du contexte socialiste de réappropriation urbaine (1979–1989) en droits de propriété." PhD diss., Institut français d'urbanisme.

– 2010. "Le Kampuchea démocratique: L'illusion d'une révolution sans ville." In *Antiurbain: Origines et conséquences de l'urbaphobie*, edited by Joëlle S. Cavin and Bernard Marchand, 233–47. Lausanne, Switzerland: Presse Polytechniques et Universitaires Romandes.

Carter, Francis W. 1979. "Prague and Sofia: An Analysis of Their Changing Internal City Structure." In *The Socialist City: Spatial Structure and Urban Policy*, edited by Richard A. French and Ian Hamilton, 425–60. Chichester, UK: John Wiley and Sons.

Castells, Manuel. 2010. "Globalisation, Networking, Urbanisation: Reflections on the Spatial Dynamics of the Information Age." *Urban Studies* 47, no. 13: 2737–45. https://doi.org/10.1177/0042098010377365

Chandler, David P. 1991. *The Tragedy of Cambodian History: Politics, War, and Revolution since 1945*. New Haven, CT: Yale University Press.

Czepczyński, Mariusz. 2008. *Cultural Landscape of Post-Socialist Cities: Representation of Power and Needs*. Aldershot, UK: Ashgate.

Dagens, Bruneau. 2008. "Angkor, instrument politique: Avant, avec et après le protectorat." In *Angkor VIIIe–XXIe siècle. Mémoire et identité khmères*, edited by Hughes Tertrais, 94–107. Paris, France: Autrement.

De Vienne, Marie-Sybille. 2012. "Le Cambodge entre réalité économique et fiction statistique (1990-2010)." *Péninsule* 64: 197–218.

Diener, Alexander C., and Joshua Hagen. 2013. "From Socialist to Post-Socialist Cities: Narrating the Nation through Urban Space." *Nationalities Papers* 41: 487–514. https://doi.org/10.1080/00905992.2013.768217

Edwards, Penny. 2007. *Cambodge: The Cultivation of a Nation, 1860–1945*. Honolulu: University of Hawaii Press.

Fauveaud, Gabriel. 2014a. "Mutations of Real Estate Actors' Strategies and Modes of Capital Appropriation in Contemporary Phnom Penh." *Urban Studies* 51, no. 16: 3479–94. https://doi.org/10.1177/0042098014552767

– 2014b. "Phnom Penh ou l'ordre métropolitain: Polices, pouvoirs et terri-
toires." *EchoGéo [Online]* 28. https://doi.org/10.4000/echogeo.13807
– 2015. *La production des espaces urbains à Phnom Penh: Pour une géographie
sociale de l'immobilier.* Paris, France: Les Publications de la Sorbonne.
– 2016. "Residential Enclosure, Power and Relationality: Rethinking
Sociopolitical Relations in Southeast Asian Cities." *International Journal of
Urban and Regional Research* 40, no. 4: 849–65. https://doi.org/10.1111
/1468-2427.12433
Fisher, Jack C. 1962. "Planning the City of Socialist Man." *Journal of the
American Institute of Planners* 28, no. 4: 251–65. https://doi.org/10.1080
/01944366208979451
French, Richard A., and Ian Hamilton. 1979. "Is There a Socialist City?" In *The
Socialist City: Spatial Structure and Urban Policy,* edited by Richard A. French
and Ian Hamilton, 1–22. Chichester, UK: John Wiley and Sons.
Gaubatz, Piper. 1999. "China's Urban Transformation: Patterns and Processes
of Morphological Change in Beijing, Shanghai and Guangzhou." *Urban
Studies* 36, no. 9: 1495–521. https://doi.org/10.1080/0042098992890
Geoffray, Marie-Laure. 2013. "La Havane après 1989: Vers une ville post-
socialiste?" In *Ségrégation et fragmentation dans les métropoles: Perspectives inter-
nationales,* edited by Marion Carrel, Paul Cary, and Jean-Michel Wachsberger,
155–70. Villeneuve-d'Ascq, France: Presses Universitaires du Septentrion.
Gerles, François. 2008. "L'économie cambodgienne." In *Cambodge contempo-
rain,* edited by Alain Forest, 189–256. Paris, France: Les Indes Savantes/
IRASEC.
Girling, John. 1971. "Cambodia and the Sihanouk Myths." Paper presented
at a seminar for the Institute of Southeast Asian Studies, University of
Singapore, 31 March.
Goldblum, Charles, and Manuelle Franck. 2007. "Les villes aux marges de la
métropolisation en Asie du Sud-Est." *L'espace géographique* 36, no. 3: 229–36.
https://doi.org/10.3917/eg.363.0229
Grant Ross, Helen, and Darryl Collins. 2006. *Building Cambodia: "New Khmer
Architecture" 1953–1970.* Bangkok, Thailand: Key Publisher Company.
Groslier, Berbard P. 1985. "L'image d'Angkor dans la conscience Khmère."
Seksa Khmer 8: 5–30.
Hirt, Sonia A. 2012. *Iron Curtains: Gates, Suburbs, and Privatization of Space in the
Post-Socialist City.* Chichester, UK: Wiley & Sons
Hirt, Sonia, and Mina Petrović. 2011. "The Belgrade Wall: The Proliferation
of Gated Housing in the Serbian Capital after Socialism." *International
Journal of Urban and Regional Research* 35, no. 4: 753–77. https://doi.org
/10.1111/j.1468-2427.2011.01056.x
Hogan, Trevor, Tim Bunnell, Choon-Piew Pow, Eka Permanasaric, and Sirat
Morshidid. 2012. "Asian Urbanisms and the Privatization of Cities." *Cities*
29, no. 1: 59–63. https://doi.org/10.1016/j.cities.2011.01.001

Lamant, Pierre-Lucien. 1991. "La création d'une capitale par le pouvoir colonial: Phnom Penh." In *Péninsule Indochinoise – Études urbaines*, edited by Pierre-Bernard Lafont, 59–102. Paris, France: L'Harmattan.

Leclère, Ademard. 1904. "La fête des eaux à Phnom Penh." *Bulletin de l'École française d'Extrême-Orient* 4, no. 1: 120–30. https://doi.org/10.3406/befeo.1904.1298

Lin, George C.S. 2004. "Toward a Post-Socialist City? Economic Tertiarization and Urban Reformation in the Guangzhou Metropolis, China." *Eurasian Geography and Economics* 45, no. 1: 18–44. https://doi.org/10.2747/1538-7216.45.1.18

Lussault, Michel. 2007. *L'homme spatial: La construction sociale de l'espace humain.* Paris, France: Seuil.

Molyvann, Vann. 2003. *Modern Khmer Cities.* Phnom Penh, Cambodia: Reyum.

Monnet, Jérôme. 2000. "Les Dimensions Symboliques de la Centralité." *Cahiers de Géographie du Québec* 44, no. 123: 399–418. https://doi.org/10.7202/022927ar

Nam, Sylvia. 2011. "Phnom Penh: From the Politics of Ruin to the Possibilities of Return." *Traditional Dwellings and Settlements Review* 23: 55–68.

Népote, Jacques, and Marie-Sybille de Vienne. 1993. *Cambodge, laboratoire d'une crise: Bilan économique et prospective.* Paris, France: CHEAM.

Osborne, Milton E. 1973. *Politics and Power in Cambodia: The Sihanouk Years.* Camberwell Victoria, Australia: Longman.

– 2008. *Phnom Penh: A Cultural and Literary History.* Oxford, UK: Signal Books.

Parnell, Susan and Jennifer Robinson. 2012. "(Re) Theorizing Cities from the Global South: Looking beyond Neoliberalism." *Urban Geography* 33, no. 4: 593–617.

Percival, Tom, and Paul Waley. 2012. "Articulating Intra-Asian Urbanism: The Production of Satellite Cities in Phnom Penh." *Urban Studies* 49, no. 13: 2873–88. https://doi.org/10.1177/0042098012452461

Peyvel, Emmanuelle. 2013. "Reshaping of Post-Socialist Ho Chí Minh City: An Approach through Leisure and Urban Practices." Paper presented at the annual meeting of the Association of American Geographers, Los Angeles, 10 April.

Phoeun, Mak. 1991. "Le phénomène urbain dans le Cambodge post-angkorien." In *Péninsule Indochinoise – Études Urbaines*, edited by Pierre-Bernard Lafont, 39–57. Paris, France: L'Harmattan.

Pierdet, Céline. 2005. "La symbolique de l'eau dans la culture cambodgienne: Fête des eaux et projets urbains à Phnom Penh." *Géographie et Culture* 56: 5–22.

Reyum. 2001. *Culture of Independence: An Introduction to Cambodian Arts and Culture in the 1950s and 1960s.* Phnom Penh, Cambodia: Reyum.

Rimmer, Peter James, and Howard Dick. 2009. *The City in Southeast Asia: Patterns, Processes and Policy.* Honolulu: University of Hawai'i Press.

Roy, Ananya. 2015. "Who's Afraid of Postcolonial Theory?" *International Journal of Urban and Regional Research* 40, no. 1: 200–9. https://doi.org/10.1111/1468-2427.12274

Scarpaci, Joseph L. 2000. "On the Transformation of Socialist Cities." *Urban Geography* 21, no. 8: 659–69. https://doi.org/10.2747/0272-3638.21.8.659

Shatkin, Gavin. 1998. "'Fourth World' Cities in the Global Economy: The Case of Phnom Penh, Cambodia." *International Journal of Urban and Regional Research* 22, no. 3: 378–93. https://doi.org/10.1111/1468-2427.00147

Sheppard, Eric. 2000. "Socialist Cities?" *Urban Geography* 21, no. 8: 758–63. https://doi.org/10.2747/0272-3638.21.8.758

Simone, AbdouMaliq. 2008. "The Politics of the Possible: Making Urban Life in Phnom Penh." *Singapore Journal of Tropical Geography* 29, no. 2: 186–204. https://doi.org/10.1111/j.1467-9493.2008.00328.x

Sliwinski Marek. 1995. *Le génocide Khmer Rouge, une analyse démographique.* Paris, France: L'Harmattan.

Springer, Simon. 2010. *Cambodia's Neoliberal Order: Violence, Authoritarianism, and the Contestation of Public Space.* London, UK: Routledge.

– 2016. "Homelessness in Cambodia: The Terror of Gentrification." In *The Handbook of Contemporary Cambodia*, edited by Kathrine Brickell and Simon Springer, 234–44. London, UK: Routledge.

Stanilov, Kiril. 2007. "Taking Stock of Post-Socialist Urban Development: A Recapitulation." In *The Post-Socialist City*, edited by Kiril Stanilov, 3–17. Dordrecht, The Netherlands: Springer.

Sýkora, Luděk. 1999. "Changes in the Internal Spatial Structure of Post-Communist Prague." *GeoJournal* 49: 79–89. https://doi.org/10.1023/a:1007076000411

Tsenkova, Sasha. 2006. "Beyond Transitions: Understanding Urban Change in Post-Socialist Cities." In *The Urban Mosaic of Post-Socialist Europe: Space, Institutions and Policy*, edited by Sasha Tsenkova and Zorica Nedovic-Budic, 21–49. New York, NY: Physica-Verlag Heidelberg.

Tyner, James A., Samuel Henkin, Savina Sirik, and Sokvisal Kimsroy. 2014. "Phnom Penh during the Cambodian Genocide: A Case of Selective Urbicide." *Environment and Planning A: Economy and Space* 46, no. 8: 1873–91. https://doi.org/10.1068/a130278p

Vickery, Michael T. 1999. *Cambodia 1975–1982.* Chiang Mai, Thailand: Silkworm Books.

Wright, Gwendolyn. 1987. "Tradition in the Service of Modernity: Architecture and Urbanism in French Colonial Policy, 1900–1930." *Journal of Modern History* 59, no. 3: 291–16. https://doi.org/10.1086/243186

– 1991. *The Politics of Design in French Colonial Urbanism.* Chicago, IL: University of Chicago Press.

7 Planning for "Renaissance": Vanguard Urbanism in Addis Ababa

JESSE McCLELLAND

"Long Live Proletarian Internationalism!" (ca 1980)
"Our renaissance journey will never be impeded by extremists [*sic*] fantasy!" (2012)

> –Banners at Abiyot/Mesqel Square, Addis Ababa

Across many of the world's rapidly growing cities, development is coupled with the possibilities of political legitimation. In African cities in particular, planning has been a crucial mode of both imperialism and socialist responses to imperialism, but the failures of each cast long shadows into the present. A great many cities in Africa are caught "between modernist ideas of how cities should look and work ... that sometimes make little sense, and an alternative, fluid, ambient – informal – city that is getting by on its own, if perhaps barely so" (Myers 2011, 79). Creative alternatives must reckon with this "crisis of urban development" (Pieterse 2008), and the vanguard urbanism being tried in Addis Ababa is a case in point. Here, the ruling Ethiopian People's Revolutionary Democratic Front (EPRDF) party draws much of its legitimation by addressing the historical legacies of the previous socialist regime in two respects. First, the inability of the socialist Derg regime to engage strategically with Addis Ababa's built environment fostered the very "slum" infrastructures that have since become irresistible targets of massive redevelopment. Second, while Abiyot Square was an iconic stage for socialist military processions of the past, the current regime's adaptation of the same site – now renamed Mesqel Square – also helps it to distinguish itself from the punitive urbanism of the Derg. While the "slum" and the square are sites of deepening ruling party engagements with urban space, a recent wave of urban uprisings also call into question the politics of legitimation through centralized planning.

Situating Derg Rule among African Socialisms

As Hirt (2013) and others have noted, a strength of the literature on socialist and post-socialist urbanisms has been its resistance to singular or essentialized categories. While this is partly due to historical contingencies that vary across the globe, the literature to date has almost exclusively focused on European and Asian contexts (see the international anthologies edited by Darieva, Kaschuba, and Krebs 2011; Kliems and Dmitrieva 2010; Czaplicka, Gelazis, and Ruble 2009; Stanilov 2007; Crowley and Reid 2002; Andrusz, Harloe, and Szelenyi 1996) and would do well to incorporate African and other experiences. Cities in Africa have helped to constitute the worldly geographies of anti-imperialism and geopolitics since the Cold War. While the optic of post-socialist urbanism makes for a relatively novel approach within Ethiopian urban studies, it shares much with a long-standing scholarly interest in Ethiopian legacies of modernist planning, development, and architecture (see Levin 2016; Zeleke 2010; Fuller 2007; Giorghis and Gérard 2007; Donham 1999; Ponsi 1982).

African socialisms of the 1960s and 1970s often emerged in the crucible of independence struggles against Western capitalist/imperialist rule. These struggles articulated across many scales, from the local demands of workers and peasants, to solidarity with liberated countries and the Organization of African Unity, to partnerships with the USSR, China, North Korea, Cuba, the GDR, and the Eastern bloc countries. In search of legitimation, anti-imperialist movements across Africa adapted Marxist-Leninist precepts to their own political grammars and purposes. In Senegal, Senghor's dual emphasis on socialism and negritude conceived of politics as a means of transmitting culture (Berktay 2010). In Tanzania, Nyerere hailed the village as the authentically African unit of social organization and thus imagined villagization as the path to *ujamaa*. As Mudimbe put it, "often formally brilliant, these socialisms, generally speaking, functioned and lived as texts marked by fantasies of an illusory new beginning of history" (Mudimbe 1994, 42). For many of the avowed socialist leaders in Africa contending with new and uncertain political economies, the challenges of finding order and legitimation led them to resort to authoritarian governance.

Quite apart from socialist visions that were grounded in long-standing movements against European imperialism, "Ethiopian Socialism" emerged in response to African leadership. Emperor Haile Selassie was widely hailed as an icon of non-alignment and of the Organization of African Unity. He was deposed in a coup in 1974 by the Provisional Military Administrative Council – popularly known as the Derg, or committee – roughly 120 men "hastily drawn from the various branches of the armed forces, police and territorial army" (Tareke 2008, 188). Four strains of opposition had weakened the emperor prior to the coup. First, students and leftist parties steeped in Addis Ababa's intellectual milieu agitated against Haile Selassie's monarchism and neo-feudalism, and compared them to exploitative conditions under European colonialism (Balsvik 2007; Tibebu 2008). Second, the poverty of peasants and the devastating famine of 1973 intensified demands for land redistribution and tax reform (Markakis 1981). Third, Eritrean, Oromo, Tigray, Somali, and other peoples challenged the Orthodox, Amharic-language institutions that had ruled since the late nineteenth century (Markakis 2011; James et al. 2002). Fourth, lower ranks of the military apparatus were demoralized; US support for Haile Selassie waned during the Nixon administration just as the emperor was ordering the pacification of popular uprisings in Eritrea (Zewde 2014). A revolutionary moment seemed to emerge when spontaneous uprisings began in Addis Ababa in February 1974. But even after resulting months of discontent that the emperor failed to quell, "the popular movement failed to develop any form of political organization" (Markakis and Ayele 1986, 180).

The Derg seized power in November 1974, and initially presented themselves as stewards of political stability. But they soon signalled a complete break with Ethiopia's legacy of highland imperialism by killing Haile Selassie and many of his advisers. Derg factions then debated models of governance, from selecting a new emperor, to working with radicals in popular forms of social organization, to the formation of a Leninist vanguard. Such debates caused the Soviets to consider Ethiopia's new leadership merely nationalist partisans instead of a revolutionary movement grounded in principle (Donham 1999).

In perhaps its most far-reaching reform, the Derg nationalized all land in 1975. This broke centuries of landholding regimes that had favoured highland aristocrats and the Orthodox Church, and it won the Derg a measure of popular support. Through new agricultural cooperatives and resettlement schemes, the Derg aimed to make rural production capacity the basis for their entry into global socialist trading networks (Dagne 2006). The Derg appropriated the slogan "Land to the Tiller!" from students and radicals to its own uses (Zewde 2002; Donham 1999). Peasant associations, the conscription of urban professionals in rural construction projects, and reeducation camps all would become sites of a new, punitive socialist order. While General Mengistu Haile Mariam and many others within the Derg were based in Addis Ababa, they were wary of the emergence of political opposition there, as in Asmara and other cities. New security and intelligence systems were started at the *kebele* (precinct) level of cities with support from the GDR (Dagne 2006; Zewde 2002). Local leaders recruited informants and sought to smash radical organizations of many kinds. Radical printing presses were seized and already weak labour unions were coopted. Addis Ababa's burgeoning Ethio-jazz scene was criticized as bourgeois entertainment and some of its leading artists fled into exile.

In 1977, following attempts by the Eritrean People's Revolutionary Party (EPRP) to assassinate Derg personnel in Addis Ababa, Mengistu publicly threatened all opponents with the Red Terror at a famous speech at Mesqel Square. Violence in Addis Ababa became ubiquitous, in house-to-house searches for dissidents, curfews, disappearances, and both judicial and extra-judicial killings.

> Victims of the extra-judicial killings were left by the roadside, often in prominent places with placards that condemned them as anarchists, anti-people or counter-revolutionaries. Tens of hundreds were thrown into unmarked mass graves. Rarely was a notice about their death or disappearance delivered to their relatives. It was intended to torment the living. (Gebru 2008, 196)

Tens of thousands died across the country during the Red Terror, and the fates of many from this period are still unknown today. Radical students decried these atrocities and criticized the Derg's champions in Moscow for "revisionism" and "Soviet social-imperialism" (ESUNA 1977), but these nascent opposition movements remained fractured and marginal.

The Derg only joined a socialist bloc in Africa in 1977 through the Ogaden War with Somalia. Eager to check US support for Somalia, the USSR, China, and the GDR all supplied Ethiopia with arms and Cuba even sent troops (Dagne 2006; Ponsi 1982). These measures both decided the war and slowed the emergence of separatist militias challenging the Derg's territorial control. Purges within the Derg's own ranks came soon after, with General Mengistu emerging as Ethiopia's dictator in 1978. Under him, the Derg sought to limit the influence of religious and ethnic identity, subsuming civil society under a variety of administrative controls. In the space now vacated by leftists and other opponents of the regime, a vanguard party called the Workers Party of Ethiopia was in place by 1984 and served as a prime mode of social control.

Socialist Interventions in Addis Ababa

Given the degree of coercion that followed the Red Terror, it is notable that the Derg struggled to engage strategically with the built environment of Ethiopia's largest city. Instead, they tended to challenge Haile Selassie's monarchist legacy through new monuments and the appropriation of public spectacles. The *Tiglatchin* (Our Struggle) monument and a statue of Marx done by an East German sculptor are among the Derg efforts still standing today. A statue of Lenin was long ago removed from Mesqel Square and now lies in a municipality warehouse. Ato Aklilu, a senior architect working today, recalls the Derg's military processions there:

> At least once a year, the 12th of September, we will have that parade going on in Mesqel Square and thousands of people will be told to go there and so on. So it was an important show – let's say, *space* – for that regime ... Declarations were made ... I remember so many things. Funny, sad, everything happened. There were people like Gaddafi coming for these events and I even saw Castro once, Fidel Castro watching this march.[1]

In time, Mengistu took to watching military processions from a gold-painted throne in order to embody some of Haile Selassie's regal flair (Donham 1999). Such aesthetic choices were emblematic of the regime's tendency to reproduce forms rather than to innovate their own.

While the activities of federal planning bureaucracies expanded in the Derg era, master planning in Addis Ababa was only undertaken once, in 1986. Most of that plan was conceived by Italian planners and financed by the Italian government, with limited consultation by Ethiopians. This relatively modest plan added the notions of urban renewal and a regional-scale framework for urban growth to the Ethiopian planning repertoire. The plan inspired Ethiopia's growing cadre of planners, including many of the senior urban planners working today. But like most of the city's master plans, few provisions would be implemented (Mahiteme 2007). Among the plan's few lasting contributions to the built environment were multi-storey housing blocks for government workers along the impressive Bole Road and a few strip malls in the city's old Piassa neighbourhood.

Indeed, the most profound change to Addis Ababa's built environment during the Derg years came from beyond the reach of the planning bureaucracy altogether, through the profusion of "slum" housing typologies. This was partly an unintended consequence of the nationalization of land, when a few dozen kebele administrations had become landlords for tens of thousands of rudimentary, one-storey homes at extremely small rents. Since original owners no longer had title and both local administration and new renters of the kebele homes were simply too poor to pay for improvements, much of the city's already strained housing stock decayed further. The kebele homes quickly became "a revenue drainer instead of a revenue earner for the government" (Kebbede 1987, 16). At the same time, the city's poor majority turned to self-provisioning through squatting, illegal subdivision, and crowding in existing structures for which there would never be titles (UN-HABITAT 2007). The Derg had lost a vital chance to move beyond coercion into a real claim on urban futures.

Ultimately, the Derg failed to resolve key challenges facing both rural and urban Ethiopia. Agrarian development and resettlement schemes led neither to robust trade or food security, but instead fostered the disastrous famine of 1984–5. Administration became mired down in expensive, drawn-out conflicts with separatist militias in several regions. As Soviet support crumbled in the late 1980s, Mengistu sought to preserve his regime and announced the end of Ethiopian Socialism. But on the promises of equitable governance in an authentic Marxist-Leninist state, the Tigray People's Liberation Front (TPLF) advanced on Addis Ababa. Mengistu negotiated his passage into exile in Zimbabwe, where he still lives today. By 1993, the TPLF had positioned itself firmly at the centre of a new coalition of four parties, the EPRDF, which has remained Ethiopia's ruling party since.

An Urban Renaissance of the Developmental State

Cities have taken on important new roles in national development agendas in Ethiopia and much of the rest of East Africa today, reflecting the conjoined forces of urbanization, globalization, and industrialization. The "*Abiyotawi* Democracy (Revolutionary Democracy)" advanced by Prime Ministers Meles Zenawi and Hailemariam Desalegn contained policymaking within centralized party/state structures that enlist cities in many ways. The EPRDF crafts increasingly sophisticated partnerships with foreign investors and has achieved concurrent years of double-digit economic growth (African Development Bank 2016). China underwrites many of Ethiopia's massive infrastructure projects, from dams to rail lines to highways to industrial parks. Investors from Asia, the Indian Ocean rim, and Europe compete for such contracts and fit their foreign manufacturing models to Ethiopia's supplies of cheap labour, land, and natural resources. Yet these private ventures are only part of the larger picture of Ethiopia's mixed economy; the EPRDF protects several state enterprises from competition, employs day labourers through its own micro-enterprises, and woos investors in the Ethiopian diaspora with specialized incentives. State commitments to infrastructure rescale the possibilities of development by fusing sectors together in new ways, and anticipating the emergence of urban regions. According to one planner, development in Ethiopia today rests quite literally "in concrete things" (Cherenet 2014). The EPRDF seeks

7.1 A landscape of massive redevelopment, as seen from the Office of the Master Plan. New light rail and commercial buildings, dilapidated housing, and striped fencing clear space for new construction. Stadium area, Addis Ababa. February 2015. Photo: Jesse McClelland.

legitimation through the delivery of precisely the bold, highly visible projects that previous regimes did not achieve.

As the possibilities of infrastructure grow, Ethiopia's largest cities have become stages for massive redevelopment, dramatically typifying Africa's urban revolution today.[2] If urban infrastructures were overlooked by the Derg's punitive socialism, they have become irresistible targets for the EPRDF's visions of an urban future. By demolishing thousands of the city's ubiquitous kebele homes and replacing them with state-funded condominiums and commercial properties, the EPRDF has responded to the legacies of previous, failed urban plans and produced legible signs of growing capacity.

Planners have thus emerged as key agents of urban governance. They are deeply aware of the historical import of their work, as well as the double-binds that the work of governance presents. Like many others in Ethiopia's growing educated middle class, planners hedge their political commitments against several professional pursuits. Ato Binyam is a senior planner dividing his professional life between the master plan office, academia, and his own private practice. During the Derg years, Binyam was an outstanding student with dreams of becoming a doctor. But at the Derg's direction, he soon found himself in a faraway socialist country on a government scholarship for architecture. He would return to become a planner during the Derg years and now contrasts the present moment with the time of Mengistu:

> You see, he was a dictator. He had his mechanisms down to the lowest level. But when it comes to professional issues like, for example, the planning of a city – decide on what should be done here and should be there – the [Derg] politicians never come out to discuss. Their purpose was politics ... They don't go out into these local, technical issues. *Now the political party is involved in everything*. You want to change the land use of this place as a planner because you think that is the proper land use. *They see it not from a technical perspective – they want to see it from a political perspective.* They want to see how they can gain personally and how they can gain in a group. So there is a very strong involvement in every dimension of life, you know? And this is why life is very difficult now because the political hierarchy is not limited in some spaces like it used to be during the Derg regime. It is every other one! If you want to buy a piece of bread, I'm sure sometime in the future you'll have to ask permission from the political party. [Italics added][3]

For Binyam, the technical and the political are separate domains of social action grounded in distinct kinds of knowledge. However, he notes that the EPRDF's municipal leadership does not accede to these distinctions. The political meetings and consultations to which Binyam and his colleagues are summoned steer the planning process toward the legal, economic, and political logics that foreclose a proper, technical approach to spatial problem solving. Massive redevelopment in Addis Ababa is hardly a laissez-faire model for an entrepreneurial city, but one ensuring the "political hierarchy" of the EPRDF "in every dimension of life." Such commitments across sites and scales are the hallmarks of vanguard urbanism.

Encountering Mesqel Square

The writing of history in Ethiopia has long been a "privileged site for successive states as an incontestable domain of knowledge, truth, legitimacy and national identity" (Toggia 2008, 319). But massive redevelopment

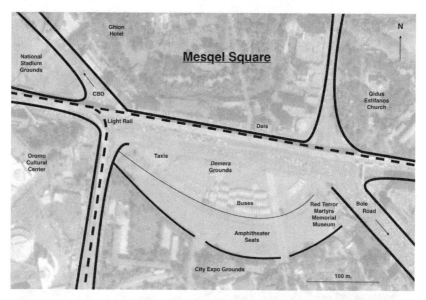

7.2 Iconic features of the Mesqel Square area, Addis Ababa. Map illustrated by Jesse McClelland using data from Google Maps, 2015.

goes one better; by inscribing state power, it draws a line against failed visions of the past. This dynamic is particularly evident in Mesqel Square, an iconic, centrally located public space in Addis Ababa that is both ordinary and spectacular. Comparable perhaps to Berlin's Alexanderplatz, an "ideological exemplar" that is also a lively space of encounter and contestation today (Weszkalnys 2007), Mesqel Square is a popular and accessible space, though use of it is mediated in a variety of ways. Most of the time, the square is a wide-open expanse offering relief from a congested city. At dawn, athletes jog together along the square's amphitheatre seating. Later in the day, Mesqel becomes a transit hub covered by queuing buses, minibuses, and taxis. Drivers, footballers, street children, chancers, young couples, and a few tourists drift across the square and linger.

Yet when Mesqel Square is activated a few days a year by assemblies and processions, it becomes a spectacular space that reduces human scale to the flat vantage point of the crowd. As the yellow Mesqel (cross) flowers begin to bloom in September, masses queue for security checks and gather for the Finding of the True Cross, the festival which leads to Orthodox New Year. Over the years, Emperor Haile Selassie, General Mengistu Haile Mariam, and Prime Minister Meles Zenawi all viewed

7.3 Thousands celebrate the Orthodox New Year with candlelight, prayer, dancing, and the traditional *demera* (bonfire), near the new light rail in Mesqel Square, September 2014. Photo: Jesse McClelland.

such events from a large dais at the edge of the square. Today, the dais has been all but obscured by the new elevated light rail train cutting across the square, an artefact of Ethio-Chinese cooperation.

Socialist legacies at Mesqel Square are now explicitly addressed at the Red Terror Martyrs Memorial Museum, which opened in 2010. Built from materials that recall the design of prison cells, this sleek and minimal museum stands adjacent to the site of an earlier archway that once proclaimed "Long Live Proletarian Internationalism!" (Interview with Aklilu, October 2014, Addis Ababa). Victims of the Red Terror of the late 1970s and their families funded the project and advocated for its construction, but it enjoyed the EPRDF's blessing and Prime Minister Meles Zenawi even laid its cornerstone. Docents there share their searing, first-hand accounts of torture, imprisonment and mourning under the Derg. They offer authoritative accounts of this chapter of history and implicitly signal to visitors that Ethiopia's current moment is a time of peace and harmony. A section of the museum was originally conceived as a discussion space for human rights events (Aklilu, interview), but the museum's board has since converted it into a small café.

Mesqel Square continues to host mass gatherings for officially sanctioned religious and political occasions. Since Prime Minister Meles Zenawi was memorialized at Mesqel Square in 2012, a banner long proclaimed, "Our renaissance journey will never be impeded by extremists [sic] fantasy!" Here the category of "extremists" lumps together terrorists and the party's political critics, whether at home or in diaspora. The stakes of that message were vividly underscored in April 2015, when so-called Islamic State militants released videos purporting to show the killing of thirty Ethiopian migrants on a Libyan beach. In response, the Ethiopian government called for a mass assembly at Mesqel Square to facilitate public mourning, to show solidarity with Ethiopians in diaspora, and to advance a unified political program ahead of the following month's national elections. Tens of thousands assembled in one of the largest public gatherings in Ethiopia since the end of socialist rule. Yet while it may have been "extremists' fantasy" that concerned the EPRDF, some of the mourners assembled were nonetheless party critics. With several opposition leaders, bloggers, and journalists behind bars in the run-up to the national elections of 2015, the assembly allowed opposition activists a rare moment of visibility and proximity to the new prime minister, Hailemariam Desalegn, as he spoke. When the event drew to a close, simmering tensions in the crowd broke open and many defied orders to disperse. Hundreds of police scrambled to clear the square with truncheons and tear gas, and detained twenty-five "ringleaders" from among the crowd. One demonstrator blamed the ruling party for the deaths of the Ethiopian migrants, claiming "It's poverty making people seek work elsewhere" (Davison 2015).

Such divergent expressions at a public assembly – of mourning, national unity, state violence, and demands for accountability – underscore the ways in which citizenship is inscribed and contested in urban space. The limits of the EPRDF's reach were arguably exposed both by the killings of Ethiopian migrants in Libya and the disruption of party interpretations of these deaths in Mesqel Square. Yet EPRDF visions of a "renaissance journey" continue to mobilize bodies in the street, in diverse sites of a globalizing labour market, and at the polls. The EPRDF coalition easily retained control of federal and local governments in the elections of May 2015.

Two years on, the dramatic encounter at Mesqel Square now appears part of a longer series of political challenges to the developmental state. Urbanization and planning have been central to the evolving national economic plans of the party, and in many quarters they have been popular expressions of improvement. However, in 2014, a burgeoning youth movement in the Oromia region began to demonstrate against the EPRDF's exclusionary models of development. In particular, they questioned the expansion of Addis Ababa, which they understood as annexation of the Oromia region that totally surrounds the city. In 2015 and 2016, the Qeeroo movement spread to many of Ethiopia's towns and cities in a few other regions, with roadblocks, clashes with police, the deaths of hundreds, and the detainment of tens of thousands (Human Rights Watch 2016). While the revision process for the Addis Ababa Master Plan was expected to have produced a revised plan in 2015, the document was shelved amid the uprising, and is not likely to be published. Amid rumors of reform both inside and outside the EPRDF, Prime Minister Hailemariam Desalegn announced a national state of emergency in October of 2016. Another urban planning round for Addis Ababa has since been convened, on the promise of more modest plans for urban expansion.

Conclusion

Ethiopia's purported national "renaissance" requires theorization of new means of legitimation basic to planning. The previous Derg regime relied on a disciplinary mode of rule, from coercive violence to an agrarian bias that undercut urban services and infrastructures in many ways. Today's ruling EPRDF party relies on many forms of violence as well. Yet its search for legitimation is tied to the production of bold and highly visible urban forms and infrastructures, particularly within cities. Planners work implicitly to address legacies of the Derg era by drawing a line against the "slum" infrastructures the city has inherited. They also work explicitly to juxtapose Mesqel Square against the Derg's campaign of Red Terror. As Ethiopian planners curate and refine existing planning practices and imagine visions for the future, they necessarily confront the ongoing legacies of socialism at the urban scale. Yet as the recent popular uprisings have hastened some reforms and the press of informal settlements continues, planning rounds have come in for rare and historic scrutiny. The urban revolution emerging in Ethiopia points to real possibilities for social justice and political change, but the terms of consent within that process are not altogether clear.

The lens of post-socialist urbanism is a generative but largely underutilized approach to the new African urbanisms flourishing today. However, a key ontological question awaits proponents of post-socialist urbanisms who want to push beyond essentialist accounts. Can this optic help in the indexing of actually existing continental and global situations, or is it an overdetermined approach bound to refer to presumed "authentic" and fully realized socialist projects? As an apparently less likely case of post-socialist urbanism, Addis Ababa is the sort of city that may overturn assumptions about what constitutes socialist projects at the urban scale, and what may characterize such projects if they are to thrive in a more worldly, uncertain, and heterodox future.

NOTES

1 Interview with Ato Aklilu, October 2014, Addis Ababa.
2 While Ethiopia's national population skews rural by African and global standards with perhaps 83 per cent of people engaged in agrarian liveli-hoods (Daniel 2011), Ethiopia's urbanization rate (2.34 per cent in 2010–15; 2.25 per cent projected in 2015–20) stands at twice the continental average (UNDESA 2014). With a population of 2.98 million in 2011, Addis Ababa was home to roughly a fifth of Ethiopia's urban population (UN-HABITAT 2014). The city may grow to 8 million by 2040 (UN, cited in Davison 2015).
3 Interview with Ato Binyam, December 2014, Addis Ababa.

REFERENCES

African Development Bank. 2016. *African Economic Outlook 2016: Sustainable Cities and Structural Transformation*. Abidjan, Ivory Coast: Author.

Andrusz, Gregoly, Michael Harloe, and Ivan Szelenyi, eds. 1996. *Cities after Socialism: Urban and Regional Change and Conflict in Post-Socialist Societies*. Hoboken, NJ: Wiley-Blackwell.

Balsvik, Randi Ronning. 2007. *The Quest for Expression: The State and the University in Ethiopia under Three Regimes, 1952–2005*. Addis Ababa, Ethiopia: Addis Ababa University Press.

Berktay, Aslı. 2010. "Negritude and African Socialism: Rhetorical Devices for Overcoming Social Divides." *Third Text* 24, no. 2: 205–14.

Cherenet, Zegeye. 2014. Keynote lecture presented at the Symposium on Climate Adapted Urban Infrastructure, Goethe-Institut, Addis Ababa, Ethiopia, 15 December.

Crowley, David, and Susan E. Reid, eds. 2002. *Socialist Spaces: Sites of Everyday Life in the Eastern Bloc*. Oxford, UK: Berg.

Czaplicka, John, Nida Gelazis, and Blair Ruble, eds. 2009. *Cities after the Fall of Communism: Reshaping Cultural Landscapes and European Identity*. Baltimore, MD: The Johns Hopkins University Press.

Dagne, Haile Gabriel. 2006. *The Commitment of the German Democratic Republic in Ethiopia: A Study Based on Ethiopian Sources*. Berlin, Germany: Lit Verlag.

Darieva, Tsypylma, Wolfgang Kaschuba, and Melanie Krebs, eds. 2011. *Urban Spaces after Socialism: Ethnographies of Public Spaces in Eurasian Cities.* Frankfurt, Germany: Campus Verlag.

Davison, William. 2015. "Islamic State Murders Spur Protests in Ethiopia's Capital." *Bloomberg News,* 22 April. http://www.bloomberg.com /news/articles/2015-04-22/ islamic-state-killings-spur-official-protest-in-ethiopia-capital

Donham, Donald L. 1999. *Marxist Modern: An Ethnographic History of the Ethiopian Revolution.* Berkeley: University of California Press.

Ethiopian Students Union in North America. 1977. "Oppose the Counter-Revolutionary Meddling of the Soviet Revisionist Renegade Clique in the Ethiopian Revolution (Part 1)." *Combat* 6, no. 2: n.p.

Fuller, Mia. 2007. *Moderns Abroad: Architecture, Cities and Italian Imperialism.* New York, NY: Routledge.

Gebreamanuel, Daniel Behailu. 2011. "Land Use Legislation in Ethiopia: A Human Rights and Environment Based Analysis." *Jimma University Journal of Law* 3, no. 2: 1–32.

Giorghis, Fasil and Denis Gérard. 2007. *Addis Ababa 1886–1941: The City and Its Architectural Heritage.* Addis Ababa, Ethiopia: Shama.

Hirt, Sonia. 2013. "Whatever Happened to the (Post)Socialist City?" *Cities* 32: S29–S38. https://doi.org/10.1016/j.cities.2013.04.010

Human Rights Watch. 2016. *Such a Brutal Crackdown: Killings and Arrests in Response to Ethiopia's Oromo Protests.* 15 June. https://www.hrw.org/report /2016/06/15/such-brutal-crackdown/killings-and-arrests-response -ethiopias-oromo-protests

James, Wendy, Eisei Kurimoto, Donald L. Donham, and Alessandro Triulzi, eds. 2002. *Remapping Ethiopia: Socialism and After.* London, UK: James Currey Press.

Kebbede, Girma. 1987. "State Capitalism and Development: The Case of Ethiopia." *The Journal of Developing Areas* 22, no. 1: 1–24.

Kliems, Alfrun, and Marina Dmitrieva, eds. 2010. *The Post-Socialist City: Continuity and Change in Urban Space and Imagery.* Berlin, Germany: Jovis Verlag.

Levin, Ayala. 2016. "Haile Selassie's Imperial Modernity: Expatriate Architects and the Shaping of Addis Ababa." *Journal of the Society of Architectural Historians* 75, no. 4 447–68. https://doi.org/10.1525/jsah.2016.75.4.447

Mahiteme, Yirgalem. 2007. "Carrying the Burden of Long-Term Ineffective Urban Planning: An Overview of Addis Ababa's Successive Master Plans and Their Implications on the Growth of the City" [working paper]. Trondheim, Norway: Norges teknisk-Naturvitenskapelige Universitet.

Markakis, John. 1981. "The Military State and Ethiopia's Path to 'Socialism'" *Review of African Political Economy* 8, no. 21: 7–25. https://doi.org/10.1080 /03056248108703464

– 2011. *Ethiopia: The Last Two Frontiers*. London, UK: James Currey Press.

Markakis, John, and Nega Ayele. 1986. *Class and Revolution in Ethiopia*. Trenton, NJ: The Red Sea Press.

Mudimbe, V.Y. 1994. *The Idea of Africa*. Bloomington: Indiana University Press.

Myers, Garth. 2011. *African Cities: Alternative Visions of Urban Theory and Practice*. London, UK: Zed Books.

Pieterse, Edgar. 2008. *City Futures: Confronting the Crisis of Urban Development*. London, UK: Zed Press.

Ponsi, Francis Thomas. 1982. "Ethiopia's Development Planning Experience (1957–1979): A Case Study of Development Theories and Activities." PhD diss., State University of New York at Buffalo.

Stanilov, Kiril, ed. 2007. *The Post-Socialist City: Urban Form and Space Transformations in Central and Eastern Europe after Socialism*. Dordrecht, The Netherlands: Springer.

Tareke, Gebru. 2008. "The Red Terror in Ethiopia: A Historical Aberration." *Journal of Developing Societies* 24, no. 2: 183–206. https://doi.org/10.1177/0169796x0802400205

Tibebu, Teshale. 2008. "Modernity, Eurocentrism, and Radical Politics in Ethiopia, 1961–1991" *African Identities* 6, no. 4: 345–71. https://doi.org/10.1080/14725840802417885

Toggia, Pietro. 2008. "History Writing as a State Ideological Project in Ethiopia." *African Identities* 6, no. 4: 319–43. https://doi.org/10.1080/14725840802417869

United Nations Department of Economic and Social Affairs (UNDESA). 2014. *World Urbanization Prospects: The 2014 Revision [Highlights]*. UNDESA Population Division. http://esa.un.org/unpd/wup/highlights/wup2014-highlights.pdf

United Nations Human Settlements Programme (UN-HABITAT). 2007. *Situation Analysis of Informal Settlements in Addis Ababa*. Nairobi, Kenya: Author.

– 2014. *The State of African Cities 2014: Re-imagining Sustainable Urban Transitions*. Nairobi, Kenya: Author.

Weszkalnys, Gisa. 2007. "The Disintegration of a Socialist Exemplar: Discourses on Urban Disorder in Alexanderplatz, Berlin." *Space and Culture* 10, no. 2: 207–30. https://doi.org/10.1177/1206331206298552

Zeleke, Elleni Centime. 2010. "Addis Ababa as Modernist Ruin" *Callaloo* 33, no. 1: 117–35.

Zewde, Bahru. 2002. *A History of Modern Ethiopia, 1855–1991*. 2nd ed. Addis Ababa, Ethiopia: Addis Ababa University Press.

– 2014. *The Quest for Socialist Utopia: The Ethiopian Student Movement c. 1960–1974*. Addis Ababa, Ethiopia: Addis Ababa University Press.

8 Recuperate, Recycle, Reuse: Adaptive Solutions for the Socialist Architecture of Bucharest

LAURA VISAN

In 1984 the end of the Nicolae Ceauşescu regime seemed like a science-fiction story to most Romanians, and the future was just "that tiny black thing that knocks at your door," as a popular joke of the era went. From Ceauşescu's side, the future looked bright: it was already planned that Casa Republicii (The House of the Republic), known today as Palatul Parlamentului (The Parliament Palace), and the new administrative centre surrounding it would be completed by 1990, along with major infrastructure works – the extension of the subway network and the construction of several underground passages under the busiest roads. It was also projected that 95 per cent of Bucharest's population would live in new apartment buildings.[1] All these would give the city "a new face, worthy of the Capital of Socialist Romania," transforming it "into an architectural model for all cities of Romania."[2]

However, the impossible happened in December 1989: the Ceauşescu regime ended a month after the fourteenth national congress of the Romanian Communist Party had acclaimed its vigour and the enthusiastic endorsement of all Romanians. After a short breath of post-socialist enthusiasm, the new state authorities were faced with the difficult task of finding new uses for the architectural landmarks of the Ceauşescu era. Although some of these constructions were not even finished, nobody seemed to mind. Ignoring them was another symbolic way to part with a repressive regime that had lasted a quarter of a century, from 1964 to 22 December 1989.

This chapter will analyse the adaptive reuse process of two categories of iconic buildings from Bucharest's socialist architecture:

1. The edifices, represented by Casa Republicii (The House of the Republic) – renamed Casa Poporului after the fall of the Communist

regime (The House of the People) and later Palatul Parlamentului (The Parliament Palace) – and by Casa Radio, and

2. The so-called hunger circuses (buildings originally intended for rationing that ended up being evidence of scarcity).

In what follows, I argue that these constructions illustrate the key principle of Ceaușescu's architectural vision for Bucharest: replacing pre-socialist remnants with monumental constructions influenced by North Korean architecture as a symbol of his political power. The second part of the chapter will present the post-1989 controversies regarding the reuse of these buildings, whose initial purpose had become obsolete. They had different fates in post-socialism: while most hunger circuses became attractive targets for real estate speculation and were transformed into shopping malls, the edifices had a more complicated trajectory, due primarily to their dimensions.

Space and Power

The influence of political regimes on urban spaces has been richly documented (Lefebvre 1991; Zukin 1993; Czepczyński 2008; Light and Young 2010; Diener and Hagen 2013). The materialization of political ideology at the level of cityscapes was particularly salient in the case of socialist societies, where it "permeated ... the most intimate spaces of the everyday with ideological meaning" (Crowley and Reid 2002, 3). Two interdependent features differentiate socialist architecture from other models: a symbolic erasure – the determination to replace all remnants of pre-socialism with *new* architecture – and the preference for "over-monumentalism ... grandiose urban spaces (plazas and triumphal axes) ... [aimed at symbolizing] political achievements, power and prestige" (Cavalcanti 1997, 82; see also Danta 1993; Czepczyński 2008, 65–71; Hirt 2006, 472; Ioan 2007, 303; O'Neill 2009, 93; Hirt 2013, 32). The edifices and the hunger circuses, two categories of iconic buildings in Ceaușescu's Bucharest, aptly illustrate both these features of socialist architecture.

The determination to eliminate the landmarks of pre-socialist architecture and to give Bucharest a new, monumental "face" became particularly visible starting in the 1970s and overlapped with a shift from an apparent political thaw to hardline socialism. In 1964, at the beginning of his presidential term, Ceaușescu was regarded favourably by Western leaders, thanks to his alleged independence from the orbit of the Soviet Union. Ceaușescu reached the peak of his popularity on 21 August 1968, when he was the only political leader from Eastern Europe to oppose the invasion of Czechoslovakia by the military forces of the Warsaw treaty. However, after a visit to North Korea and China in 1970, Ceauşescu decided to submit Romania to a Cultural Revolution model. From a "maverick" (Tismaneanu 2003, 187), he rapidly transformed into a dictator who "treated Romania as his personal domain" (Linz and Stepan 1996, 347).

The Edifices

The House of the Republic

On 4 March 1977, Romania was shattered by an earthquake with a magnitude of 7.2 on the Richter scale. The south-eastern part of the country, including Bucharest, was severely hit, and 1,424 people died – at that time, the population of Bucharest was 1,807,239 (Insse.ro 2002).[3] This gave Ceauşescu the pretext to initiate an operation of "major 'urban renewal'" in Bucharest (Ioan 2006, 337). Although some of the buildings damaged by the earthquake could have been renovated, the president decided to build a new political and administrative centre for Bucharest, which would have at its core Casa Republicii and the adjacent boulevard Victoria Socialismului (The Victory of Socialism), which aimed to exceed in length and width the Champs Élysées in Paris. His ambition was to transform Bucharest into the top socialist city of the world, a high-rise haven for the "new man" who would live there (Cavalcanti 1997, 85–6; Light and Young 2010, 9; Panaitescu 2012, 174–5).

Like Stalin and Khrushchev, who demolished all buildings reminiscent of the czarist era (Andrusz 2006, 73), Ceauşescu ordered centuries-old buildings and several Orthodox monasteries to be bulldozed, in a "megalomaniac wish" to create a new centre of Bucharest that would bear the exclusive stylistic and ideological imprint of the Ceauşescu era and would prevent people from growing attached to the pre-socialist cityscape (Panaitescu 2012, 42). Bucharesters witnessed an "apotheosis of erasure" that began with "the demolition of 4,850,000 square metres – roughly the size of Venice – of the downtown area" (Ioan 2003, 254; see also Cavalcanti 1997, 95; Ioan 2007).

In a similar fashion to Stalin, Ceauşescu considered himself the greatest architect and nation builder. This image had both a literal and a symbolic meaning. Ceauşescu made frequent visits to construction sites and gave directions to architects and builders, but specialists were exasperated by his inept interventions, which they saw as an exercise of power through "decrees, official regulations or planning directives" (Cavalcanti 1997, 87; see also Panaitescu 2012, 42). At a symbolic level, the propaganda of the Romanian Communist Party rendered Ceauşescu as an "architect of renewal" and an artisan of Romania's break with the pre-Communist past. Though media accounts presented monumental architecture as a brilliant achievement of the entire Romanian people (Şincan 1988, my translation), Casa Republicii represented a notorious example of "occult architecture" because public money was directed for the private use of Nicolae Ceauşescu: "his own private House of the People, his own river, his own city, his own country" (Zahariade 2011, 120). So-called occult architecture was not accessible to ordinary Romanians (though physically visible, it was concealed from them). Though the House of the Republic was giant, and thus visible from very far away, ordinary Romanians were not even allowed to drive on the adjacent streets. There was an air of mysteriousness and ominousness that surrounded these giant buildings because no ordinary Romanians knew what was happening inside. It was an Orwellian reality. Ordinary Romanians were not even allowed to drive on the Victoria Socialismului boulevard, a restriction eliminated after the 1989 Revolution.

The construction work for Casa Republicii and the Victoria Socialismului boulevard was launched by Nicolae Ceauşescu and his wife, Elena, on 25 June 1984. The press wrote extensively about the presidential patronage of the "monumental architectural oeuvre, representing the brightest era in the history of our country."[4] A ceremony preceded the beginning of construction. The Ceauşescus signed a parchment bearing the coat of arms of the Socialist Republic of Romania, which was then placed inside a steel cylinder and deposited where the building's foundation would be, along with a few scoops of concrete, aimed to consecrate the site. Planned to be finished in 1990, the construction was intended to host Marea Adunare Naţională (The Grand National Assembly, the Communist version of a parliament), the State Council, and the Central Committee of the Romanian Communist Party.

The new administrative centre of Bucharest emulated the architecture of Pyongyang, "a city with the vocation of construction," which boasted grand boulevards, monumental architecture, mosaics of over 500 square metres each, and a new Pioneers' House with over 500 rooms (Pascal 1989, 6). The exterior monumentality of Casa Republicii was complemented by its lavish internal decorations; however, its luxurious kitsch, with kilograms of gold decorations used, represented "a spectacular misdirection of resources," considering the austerity imposed on Romanians (Light 2001, 1061). The influence of North Korean architecture differentiated Bucharest from all other Eastern European capitals. It was aimed to reflect Ceauşescu's distancing from the surrounding socialist countries, which had begun to allow the development of private economic initiative, albeit timidly,

8.1 Palatul Parlamentului (the Parliament Palace) in 2015. Photo by the author.

in response to Gorbachev's *perestroika* model. In turn, Ceauşescu remained committed to a hardline socialism, similar to the North Korean version.

Casa Radio

The construction of this monumental building began in the last years of Ceauşescu's regime, but was never finished. With a projected surface of over 150,000 square metres, the gigantic construction was intended to host the National History Museum of the Socialist Republic of Romania and would have twenty-two 1,000-square-metre meeting halls. It boasted a big balcony that would "represent the official tribune for the big demonstrations of working people from the capital, on the occasion of significant political events for our country."[5] From that balcony, Nicolae Ceauşescu, his wife, and the high-ranking officials of the

Romanian Communist Party watched the last national day parade, on 23 August 1989 (Cioroianu 2010).

The Hunger Circuses

1984 brought the first official statements about the need to impose a program of scientific nutrition in Romania, "aimed to contribute ... to a balanced fulfillment of the alimentary consumption requirements of all members of our society" (Vasilescu and Cîmpeanu 1984, 4). It was, in fact, a pretext for the rationing of food, which was already scarce on the Romanian market because most products were exported. State authorities decided to build six alimentary complexes, one in each district of Bucharest, to enact Ceauşescu's directives for so-called scientific nutrition. The regime's inclination for monumentality manifested in this case too: the first complex of this kind was planned to be "over 100 metres in length, 60 metres in width, a total surface of 17,000 square metres ... and [with] a beautiful cupola with an opening of approximately 40 metres" (Bondoc 1985, 3).

All complexes were planned to boast the same cupola roof, reminiscent of the iconic building of the Bucharest Circus, which made people mockingly call them "the hunger circuses." Romanians have been dealing with a chronic shortage of food since the early 1980s, when "you couldn't find sugar or oil. The potatoes you bought at the grocer's were the size of a nut. They sold us two to three kilos, which was supposed to be enough for one family. However, we could find frozen fish everywhere" (Martor 2002).

The opening of the first complex was lavish, with an abundance of food displayed to delight Ceauşescu, who attended the event: "It was like at the Harvest Day, culinary exhibitions and [Ceauşescu's] working visits, all in one place. All other food stores were bare empty." Unsurprisingly, the wonder did not last. After a short while, the complex was empty: "Between the concrete giant and a regular food store there was no difference. Without food, the complex was useless and depressing" (Barbulescu 2009, my translation).

One of the most important functions of these complexes was as collective canteens that could host hundreds of people. Under the pretence of care for people's health, the Ceauşescu regime was in fact enforcing a biopower exercise reminiscent of Orwell's *1984* through the "centralization and regularization of both the purchase and consumption of food" (Light and Young 2010, 8). Gradually, "cooking in every family would have faded away" (Martor 2002), and the people would have been compelled to dine together in the new complexes. By 1989, just one of these constructions was finished, though the others were close to completion.

Although propaganda capitalized on the pride of living in the soon-to-be most modern socialist capital of the world, Bucharesters remained indifferent, if not hostile, towards the new architecture, which was devoid of any cultural, affective, or spiritual significance for most of them. In an unexpected twist of fate, some of these grandiose constructions became spaces of dissent against the Communist regime that had created them (Diener and Hagen 2013, 495–6).

Post-Socialist Changes

After the euphoria of the 1989 Revolution and the overthrow of the Ceauşescu regime, Romania entered a phase of transition that "abound[ed] in ambiguities, paradoxes, regression attempts, identity losses, frauds and acts of imposture" (Anghelescu 2008, 5, my translation). The socialist construction sites were quickly abandoned, and the media shifted from acclaiming them as the pride of the city and the core of socialist urban renewal to decrying the health hazards of these spots: "In Bucharest the air usually has a greyish colour ... we all know where it comes from: demolitions, constructions, dirty streets" (Marin 1990a, 2). The gleaming capital of Ceauşescu's Romania gave way to a dystopian city reminiscent of Dickens's London: "Splaiul Unirii, which we boast of, has been exclusively grabbed by pedlars, rattlers, gypsy women selling old junk, thieves, gangsters, beggars, popcorn-makers, barbeque-makers, traffickers, drunkards, stray dogs, and foreigners who fool around as they wouldn't dare to do at home" (Gavrilescu 1990, 1).

The transition from socialism to capitalism also brought a significant paradigm change: the shift from "commonality" to a "post-public" domain (Hirt 2012, 47). During socialism, the state could construct monumental buildings and ample public spaces because it had a monopoly on land. Things changed after 1990 when, in line with the mechanisms of a market economy, land became a commodity and an object of real estate speculation (Hirt 2013, 29; Wring 2013).

From Casa Republicii to Casa Poporului

One of the first decisions the new government of Romania made was to rename the Casa Republicii (The House of the Republic) the Casa Poporului (The House of the People) in an act of symbolic restitution.

Soon the building was opened as a museum for the public, though the construction work had not been finished.

In 1990 the Romanian Union of Architects organized a debate about the "Casa Poporului and the Reality of the City," but architects and public authorities took opposing positions. Architects favoured a more radical set of solutions, all centred on symbolic acts of recuperation. Some of them suggested rebuilding the old streets that were demolished before 1989, which would now cross through Casa Poporului, thus downgrading its role. Some suggested recreating the Arsenal Hill, which had been demolished during the process of construction – the building could be buried beneath the hill like a tomb. Another suggestion would be to open the building up – to "bring the city into the house" by piercing the construction and opening the space for commercial purposes. The chief architect of Bucharest dismissed all the proposals, suggesting that architects should provide solutions for Casa Poporului only, but not for its surrounding space. The debate was emblematic of the early 1990s desire to break away from the socialist past – a desire that coexisted with a conservative, past-oriented approach (Stroe-Brumariu 1991, 4).

At a similar debate organized in 1991, the architects and officials remained irreconcilable in their positions. In the meantime, the government had granted the parliament the right to use Casa Poporului. The lower chamber of the Romanian Parliament and its commissioners rushed to move into the completed area of the building, while the senate, the upper chamber, waited. Commenting on their haste, Alexandru Beldiman, the president of the Union of Architects, maintained: "In a moral order of things, I consider that in a building where Ceaușescu wanted to place power, it is a mistake to promote the same thing after the 1989 Revolution." He recommended finding another function for the building that would both cover its expenses, and generate revenue. However, a parliament member who attended the meeting disagreed, pointing that "all the world wonders were constructed under dictatorships," and that Casa Poporului was "intended for the representatives of power, so this is where power should remain" (Stroe-Brumariu 1991, 4).

The solutions proposed by architects converged in the direction of a symbolic restitution, to "honor those who lost their homes and churches, who died because of trauma caused by demolition or by forced labor at the House of the Republic site itself, and who were unaccounted for after 1989" (cited in Tsenkova 2008, 343). Architects also recommended that an international contest be organized to seek solutions for the Casa Poporului. This would crystallize in the Bucharest 2000 contest, another initiative "of notable echo, but with relatively scarce pragmatic effects" (Ioan 2003, 78).

Architects' views diverged from the "non-specialists" who visited Casa Poporului after it had been opened as a museum. An architect visiting from Norway wrote in the museum book that the construction should be demolished, covered by mud, and transformed into a symbol of evil that is destroyed and buried. "This gentleman should be ashamed of himself," the next visitor, a Romanian, wrote. In a similar vein, many Romanians considered that the building should be finalized as a tribute to the years of toiling and deprivations experienced by Romanians (Sălăgean 1991).

In 1994 Casa Poporului hosted the meetings of the Crans Montana economic forum, and in 1995 the name of the building was changed to Palatul Parlamentului (The Parliament Palace). In 2005 the upper chamber, the senate, also moved here. Had the presidential office been moved here as well, "Ceauşescu's testament would have been fully fulfilled," Augustin Ioan (2007, 305) bitterly remarked.

The pressures of neoliberalism could not have spared the Palatul Parlamentului. Local authorities attempted to commodify its construction by transforming it into a profitable establishment. Exotic solutions for adaptive reuse suggested that a part of the construction could become a Dracula theme park with Michael Jackson as an endorser, a luxury mall financed by Middle Eastern investors, a giant casino, or a hotel with a golf course owned by the Parliament (Ioan 2007, 307; Ioan cited in Tsenkova 2008, 339; Light and Young 2010, 10), but none of these materialized.

In time, the Palatul Parlamentului became a tourist brand of Bucharest. It is a common trait of former socialist cities to reuse iconic buildings of the totalitarian regime, transforming them into "curiosities" for amateurs of "commie tourism" (Diener and Hagen 2013, 489). Bucharesters do not yet love – or even like – the building, but they are growing to accept it. Early in 2015 the Association for the Promotion and Touristic Development of Bucharest announced that the capital city would be advertised abroad with the Palatul Parlamentului and the Old Centre of Bucharest (DIGI 24 2015). It was an interesting choice to bring together a landmark of the Ceauşescu era, built over an old area that had been demolished, and the Old Centre. Knowing the fervency of the Communist regime to remove the traces of pre-socialism, it is probable that, in time, the Old Centre would have become a candidate for demolition had the 1989 Revolution not happened. However, choosing the Palatul Parlamentului as a symbol of tourist Bucharest appears to have been an inspired choice, at least from a Western perspective. Despite the apprehension of Bucharesters, the Palace of Parliament is ranked among the seven unknown architectural wonders of the world in a list recently issued by the BBC (Călin 2015).

Casa Radio

Soon after the 1989 Revolution, the new government proposed that the Romanian Radio Society should move into the monumental building as soon as it was completed, but the institution considered it inappropriate for its technical requirements. Though the move never happened, the unfinished building has remained in the collective memory as "Casa Radio." In the absence of any solution for reuse, the construction site of Casa Radio was abandoned, and the giant concrete frame quickly devalued because of galloping inflation.

In June 2007 it was announced that 70 per cent of the building would be demolished and replaced with a five-star hotel, offices, and commercial spaces (RL Online 2007). This radical decision was soon reversed and a new plan was revealed at the end of 2007: a consortium of real estate developers from Israel, Netherlands, and Turkey intended to transform the abandoned construction, renamed Dâmbovița Centre – from the name of the adjacent river, Dâmbovița – into a "symbol of Bucharest," a paradise of consumerism hosting "a giant Ferris-wheel like the London Eye and multicolored fountains 'as in Las Vegas'" (Boiangiu 2007, my translation). It was estimated that the project would be completed in forty months.

The initiative raised the scepticism of the Order of Architects and the non-governmental organization Salvați Bucureștiul (Save Bucharest), through its representative, Nicușor Dan. The main concerns about the new construction included increased traffic congestion, degradation of the environment in a neighbourhood where parks were already scarce, a sense of disproportion between the volume of the building and its surrounding area, and excessive commercial development (Boiangiu 2009).

Following the economic crisis that occurred in 2009, the relations between the consortium partners deteriorated and the construction work stopped. Abandoned again in 2010, the construction has been degrading. After the global financial crisis, the real estate value of the project decreased from €772,35 million in 2010 to €331,7 million in 2012 (Tamas 2010). Eventually, the real estate consortium went into financial collapse after Standard and Poor's reduced the rating of their corporate credits (Păvăloi 2013). In 2015 the Dâmbovița Centre, known by

8.2 Dâmbovița Centre (Casa Radio), still unfinished in 2015. Photo by the author.

Bucharesters as Casa Radio, is still an unfinished construction with uncertain prospects of being finalized in the future.

From Hunger Circuses to Shopping Malls

A common characteristic of post-socialist economies was the shift from heavy industrialization – no longer sustainable after the demise of Comecon (The Council for Mutual Economic Assistance, the socialist counterpart of the Organization for European Economic Co-Operation) – to a retail and service-based economy. The transformation in Romania was piecemeal, and it began with a new form of improvised commerce: small convenience stores opened in garages, in ground-level apartments, or in the hallways of apartment buildings (Tsenkova 2008, 295). In Bucharest, these small stores were pejoratively called *dughene* or, with a more aspirational term, *buticuri,* from the French *boutique.* The newly discovered private initiative manifested through "a frenzy of erecting ugly and insalubrious ramshackle kiosks of corrugated iron stolen from former state factories" (Marin 1990b, 1).

However, Romanians did not seem to mind the dubious aesthetics of the improvised stores, as long as they had access to goods they had coveted for so long: "Everything was a wonder [to us]. While before the Revolution we didn't have anything, now we had everything. It was breathtaking, and we enjoyed all breadcrumbs that were thrown to us. Stores were full – full of junk, though, but we understood only this later" (woman, forty-two-year-old designer, quoted in Popovăţ 2008, 25, my translation). The goods for small commerce were purchased through quick trips to Western Europe, Turkey, and even India, China, or Thailand. These provided the supply of cheap goods Romanians were looking for at the time. Some stores carried only products that were considered upscale in the early 1990s – chocolate bars, soft drinks, peanuts, blue jeans, or perfumes. In the absence of any market regulations, many of these stores sold their goods in foreign currency – US dollars or German marks – to avoid the rapid devaluing of the Romanian currency, the leu.

In the meantime, nobody seemed to care about the unfinished hunger circuses, which represented "unwanted symbols, physical reminders of a period of austerity and deprivation that everyone wanted to forget" (Light and Young 2010, 8). Temples of never-attained socialist cornucopia, these buildings soon began to deteriorate. Left unsupervised, the sites invited looting of the abandoned fixtures and construction materials. Ideas on how to reuse these spaces soon emerged: warehouses, parking lots, car-repair shops (Militaru 1991). The only complex that had been finished in 1989 used to sell frozen fish and shrimp chips while peasants sold meat on the sidewalk outside the building (Ionescu 1990).

The adaptive reuse solution came in 1989 when Vitan was the first hunger circus finished and converted into a shopping mall, Bucharest Mall, the first of its kind in the Romanian capital. Throughout the post-socialist space, the mushrooming mom-and-pop shops could no longer meet the demands of customers, particularly of a growing middle class that "longed for opportunities to demonstrate their newly acquired wealth, aspiring to reach Western standards of consumption" (Stanilov 2007, 88; see also Tsenkova 2008, 295). For Eastern Europeans accustomed to the frugality imposed by totalitarian regimes, malls represented "the quintessential form of contemporary consumerism ... blurring boundaries between shopping, entertaining and leisure" (Andrusz 2006, 71), and a "self-conscious symbol of capitalism as opposed to socialism" (Andrusz 2006, 85; see also Stanilov 2007, 7).

Converting the former hunger circus into a shopping mall represented the transition from socialism to capitalism: "a former 'factory' of regularised and collectivised food retailing was turned into a temple of individualised consumption" (Light and Young 2010, 6).

Romanians were delighted with the new mall, which soon became the favourite destination for shopping, walking, and being noticed. Like the other recently opened malls in Eastern Europe, Bucharest Mall was perceived as a "high status palace of consumption ... never aspir[ing] to catering for a mass market" (Andrusz 2006, 84). Many people travelled from other cities just to visit Bucharest Mall. The artesian fountain in its hallway was the talk of the town. Through its amenities and selection of shops, Bucharest Mall "set the international standards for retail and entertainment" (Bucureşti Mall 2015, my translation). However, the glory of Bucharest Mall began to fade during the mid-2000s, when another hunger circus was converted into the Plaza Romania mall. This one, too, lost some of its magic when a 300-store mall opened close by, on the platform of the former Electrical Machines Factory.

Conclusion

As this chapter has attempted to demonstrate, the hunger circuses appear to have been the most successful example of adaptive reuse in Bucharest's transition from socialism to post-socialism. Casa Poporului, which now hosts the Parliament of Romania, has had its share of exotic reuse proposals, reflecting the turbulent transition of the early 1990s. Casa Radio lays abandoned in the centre of Bucharest.

The hunger circuses have been gradually converted into shopping malls and now compete with each other in attracting the money of Romanians who, in time, have become discerning consumers. Another post-socialist irony is that they also compete with former factories and industrial platforms, the pride of the Ceauşescu regime, now also converted into shopping malls. In 2015 such constructions were still considered remnants of the Ceauşescu era by Bucharesters "who lived alongside these structures and despised them for that political order that they represented" (Light and Young 2010, 6).

It remains to be seen when the post-socialist label will be discarded – that is, when these constructions will get a life of their own, outside the socialist/post-socialist paradigm. Wishing to photograph the infamous cupolas of the former hunger circuses that are now malls, I was astonished to notice that the exterior walls had been raised to cover the cupolas, which are now visible only from the inside. The former temples of "scientific nutrition" have been contained within the opaque fortresses of consumerism, a symbolic appropriation of socialist remnants by post-socialist city forms. At this moment, the socialist and post-socialist cityscapes "coexist as layers of new development ... superimposed over the old urban fabric" (Stanilov 2007, 8), but time will tell when these two layers will conflate, and how the public perception of such constructions will change when socialism becomes a remote memory in the collective mind of Bucharest.

NOTES

1   Another projected ideal for Romania was an increase of its population to 30 million people in 2000. This would be achieved by obliging all women under forty-five to have four children each. Contraception and abortion were forbidden. In December 1989 the population of Romania was slightly over 23 million; since then, it has decreased constantly to 19,861,000 million people on 1 January 2015, according to data released by the National Statistics Institute of Romania.
2   "Cuvântarea Tovarăşul Nicolae Ceauşescu la Conferinţa Organizaţiei de Partid a Municipiului Bucuresti," *Scînteia* [The Spark], 10 November 1984. All translations from Romanian are mine.
3   According to the National Statistics Institute of Romania, the population of Bucharest was 2,067,545 in 1992 and 2,103,346 on 1 January 2015 (Insse.ro 2015).
4   "Tovarăşul Nicolae Ceauşescu, Împreună cu Tovarăşa Elena Ceauşescu, A Inaugurat Lucrările de Construcţie la "Casa Republicii" şi Bulevardul Victoria Socialismului," *Scînteia*, 26 June 1984. My translation.
5   "Vizita de Lucru a Tovarăşul Nicolae Ceauşescu, Împreună cu Tovarăşa Elena Ceauşescu, în capital," *Scînteia*, 18 August 1984. My translation.

REFERENCES

Andrusz, Gregory. 2006. "Wall and Mall: A Metaphor for Metamorphosis." In *The Urban Mosaic of Post-Socialist Europe*, edited by Sasha Tsenkova and Zorica Nedović-Budić, 71–90. Heidelberg, Germany: Physica-Verlag.

Anghelescu, Şerban. 2008. "Culorile Tranziţiei." In *Muzeul Ţăranului Român-Mărturii orale anii '90 şi bucureştenii*, edited by Şerban Anghelescu, Gabriela Cristea, Vlad Manoliu, Carmen Mihalache, and Petre Popovăţ, 5–12. Bucharest, Romania: Paideia.

Bărbulescu, Răzvan. 2009. "De la Fabrica de mâncare la Circul foamei." *Jurnalul.ro*, 18 February. http://jurnalul.ro/scinteia/jurnalul-national /de-la-fabrica-de-mancare-la-circul-foamei-319162.html#

Boiangiu, Cristina. 2007. "Capitala, admirată dintr-o roată uriaşă." *România Liberă*, 6 November. http://www.romanialibera.ro/actualitate/proiecte-locale/capitala--admirata-dintr-o-roata-uriasa-110520

– 2009. "Proiectul Dâmboviţa Center, atacat de ONG-uri." *România Liberă*, 17 February. http://www.romanialibera.ro/actualitate/proiecte-locale/proiectul-dambovita-center--atacat-de-ong-uri-146450

Bondoc, Gabriela. 1985. "Vă informăm despre: Piaţa multifuncţională Pantelimon-Delfinului." *Scînteia*, 12 May, 3.

Bucureşti Mall. 2015. "Bucureşti Mall – Primul centru comercial deschis în România." http://bucurestimall.com.ro/despre-noi

Călin, Dorina. 2015. "BBC: Palatul Parlamentului, printre cele şapte minuni arhitecturale necunoscute ale lumii." *Ziarul Financiar*, 11 February. http://www.zf.ro/zf-24/bbc-palatul-parlamentului-printre-cele-sapte-minuni-arhitecturale-necunoscute-ale-lumii-13815538

Cavalcanti, Maria de Betânia Uchôa. 1997. "Urban Reconstruction and Autocratic Regimes: Ceauşescu's Bucharest in Its Historic Context." *Planning Perspectives* 12: 71–109. https://doi.org/10.1080/026654397364780

Cioroianu, Adrian. 2010. "23 August şi Ironia Istoriei – Început cu Scandal, Sfârşit cu Butaforie." *Historia*, 17 August. http://www.historia.ro/exclusiv_web/general/articol/foto-23-august-i-ironia-istoriei-nceput-scandal-sf-r-it-butaforie

Crowley, David, and Susan E. Reid. 2002. "Socialist Spaces: Sites of Everyday Life in the Eastern Bloc." In *Socialist Spaces: Sites of Everyday Life in the Eastern Bloc*, edited by David Crowley and Susan E. Reid, 1–22. Oxford, UK: Berg.

Czepczyński, Mariusz. 2008. *Cultural Landscapes of Post-Socialist Cities – Representation of Powers and Needs*. Hampshire, UK: Ashgate.

Danta, Darrick. 1993. "Ceauşescu's Bucharest." *The Geographical Review* 83, no. 2: 170–82. https://doi.org/10.2307/215255

Diener, Alexander C., and Joshua Hagen. 2013. "From Socialist to Post-Socialist Cities: Narrating the Nation through Urban Space." *Nationalities Papers* 41, no. 4: 487–514. https://doi.org/10.1080/00905992.2013.768217

DIGI 24. 2015. "Bucureştiul va avea brand turistic şi va fi promovat în străinătate cu viaţa de noapte şi Palatul Parlamentului." *DIGI 24*, 1 May. http://www.digi24.ro/Stiri/Digi24/Actualitate/Stiri/Bucurestiul+va+avea+brand+turistic+Va+fi+promovata+viata+de+noap

Gavrilescu, Al. 1990. "Periferia din Centrul Capitalei: Ce Nemaipomenită Harababură!" *Adevărul*, 25 November, 1.

Hirt, Sonia. 2006. "Post-Socialist Urban Forms: Notes from Sofia." *Urban Geography* 27, no. 5: 464–88. https://doi.org/10.2747/0272-3638.27.5.464

– 2012. *Iron Curtains: Gates, Suburbs and Privatization of Space in the Post-Socialist City*. West Sussex, UK: Wiley.

– 2013. "Whatever Happened to the (Post)socialist City?" *Cities* 32: 29–38.

Insse.ro. 2002. "Populația la Recensamintele din Anii 1948, 1956, 1966, 1977, 1992 si 2002 – Judete si Medii." http://www.insse.ro/cms/files /RPL2002INS/vol1/tabele/t01.pdf

Insse.ro. 2015. "Populația României pe localități la 1 ianuarie 2015." http:// www.insse.ro/cms/files/publicatii/pliante%20statistice/Populatia%20 Romaniei%20pe%20localitati%20la%201%20ian%202015.pdf

Ioan, Augustin. 2003. "Bucharest – The Uncompleted Project." In *Teme ale Arhitecturii din Romania in Secolul XX,* edited by Ana-Maria Zahariade, Augustin Ioan, Mariana Celac, and Hans-Christian Maner, 253–7. Bucharest, Romania: Editura Institutului Cultural Român.

– 2006. "Urban Policies and the Politics of Public Space in Bucharest." In *The Urban Mosaic of Post-Socialist Europe,* edited by Sasha Tsenkova and Zorica Nedović-Budić, 337–48. Heidelberg, Germany: Physica-Verlag.

– 2007. "The Peculiar History of (Post) Communist Public Places and Spaces: Bucharest as a Case Study." In *The Post-Socialist City,* edited by Kiril Stanilov, 301–12. Dordrecht, The Netherlands: Springer.

Ionescu, Mihai. 1990. "Piețele Neputinței." *Adevărul,* 30 November, 2.

Lefebvre, Henri. 1991. *The Production of Space.* Translated by Donald Nicholson-Smith. Malden, MA: Blackwell.

Light, Duncan. 2001. "Facing the Future: Tourism and Identity-Building in Post-Socialist Romania." *Political Geography* 20, no. 8: 1053–74. https://doi .org/10.1016/s0962-6298(01)00044-0

Light, Duncan, and Craig Young. 2010. "Reconfiguring Socialist Urban Landscapes: The 'Left-Over' Spaces of State-Socialism in Bucharest." *Human Geographies* 4, no. 1: 5–16.

Linz, Juan J., and Alfred Stepan. 1996. *Problems of Democratic Transition and Consolidation: Southern Europe, South America, and Post-Communist Europe.* Baltimore, MD: The Johns Hopkins University Press.

Marin, Ion. 1990a. "Vă place orașul în care trăiți?" *Adevărul,* 21 March, 2.

– 1990b. "Primăria și Dughenizarea." *Adevărul,* 3 December, 1.

Martor. 2002. "Litera C." *Martor: Revista de antropologie a Muzeului Țăranului Român* 7. http://martor.memoria.ro/index.php?location=view _article&id=20&page=9

Militaru, N. 1991. "Abandonate, pe răspunderea cui?" *Adevărul,* 1–2 June, 1.

O'Neill, Bruce. 2009. "The Political Agency of Cityscapes." *Journal of Social Archaeology* 9, no. 1: 92–109. https://doi.org/10.1177/1469605308099372

Panaitescu, Alexandru. 2012. *De la Casa Scânteii la Casa Poporului, Patru decenii de arhitectură în București 1945–1989.* Bucharest, Romania: Simetria.

Pascal, Radu. 1989. "Phenian–vocația construcției; Note de drum din R.P.D. Coreeană." *Scînteia,* 22 June, 6.

Pǎvǎloi, Bogdan. 2013. "Dezvoltatorul proiectului de la Casa Radio se află în colaps financiar." *Adevărul,* 8 January. https://adevarul.ro/news

/bucuresti/dezvoltatorul-proiectului-casa-radio-afla-colaps-financiar-1
_50ebc6a956a0a6567e517d2d/index.html

Popovăţ, Petre. 2008. "Zvonuri, bârfe şi păreri." In *Muzeul Ţăranului Român-Mărturii orale anii '90 şi bucureştenii*, edited by Şerban Anghelescu, Gabriela Cristea, Vlad Manoliu, Carmen Mihalache, and Petre Popovăţ, 23–9. Bucharest, Romania: Paideia.

RL Online. 2007. "Casa Radio Va Fi Demolată." *România Liberă*, 7 June. http://www.romanialibera.ro/actualitate/eveniment/casa-radio-va-fi-demolata-97523

Sălăgean, V. 1991. "Casa Poporului (VI), Ce-i de făcut." *Adevărul*, 25 March, 3.

Şincan, Dionisie. 1988. "Bucureşti, cu faţa spre secolul XXI." *Scînteia*, 11 September.

Stanilov, Kiril. 2007. "The Restructuring of Non-Residential Uses in the Post-Socialist Metropolis." In *The Post-Socialist City*, edited by Kiril Stanilov, 73–99. Dordrecht, The Netherlands: Springer.

Stroe-Brumariu, Raluca. 1991. "Casa Poporului din nou in discuţie." *România Liberă*, 28 September, 4.

Tamaş, Cristina. 2010. "Ce se mai întâmplă pe piaţa spaţiilor comerciale?" *România Liberă*, 19 October. http://www.romanialibera.ro/economie/finante-banci/ce-se-mai-intampla-pe-piata-spatiilor-comerciale-203084

Tismaneanu, Vladimir. 2003. *Stalinism for All Seasons: A Political History of Romanian Communism*. Berkeley: University of California Press.

Tsenkova, Sasha. 2008. "Managing Change: The Comeback of Post-Socialist Cities." *Urban Research and Practice* 1, no. 3: 291–310. https://doi.org/10.1080/17535060802476525

Vasilescu, Adrian, and Alexandru Cîmpeanu. 1984. "Cronica lunii iunie – internă si internaţională." *Scînteia*, 3 July, 4.

Wring, Roxana. 2013. "Bucharest: In Search of Lost Identity." *Arhitectura* 2, no. 644: 16.

Zahariade, Ana Maria. 2011. *Arhitectura in proiectul communist: România 1944–1989*. Bucharest, Romania: Simetria.

Zukin, S. 1993. *Landscapes of Power: From Detroit to Disney World*. Berkeley: University of California Press.

9 The Paradox of Preserving Modernism: Heritage Debates at Alexanderplatz, Berlin

MARKUS KIP AND DOUGLAS YOUNG

In this chapter, we consider the debate over the future of Berlin's Alexanderplatz, the area in East Berlin that was reshaped in the 1960s to be an important central place in the *"Haupstadt der DDR."* East Berlin was the capital city of the German Democratic Republic (GDR), the country established in 1949 and known in the West as East Germany. The debate has been ongoing since German reunification in 1990, but attracted greater interest and controversy in 2013 with the Berlin building senator's announcement that she was considering granting heritage preservation status to some still remaining GDR-era buildings at Alexanderplatz that were poised for demolition. Eventually, in July 2015, two large buildings from the GDR period were listed. The arguments in the ongoing debate around preservation at Alexanderplatz differ in their emphasis on aesthetic, historical, and economic considerations. Within this spectrum, advocates for protecting GDR-era buildings focus on aesthetic and historical reasons, presenting them as testimonies to the *Ostmoderne*, or socialist-era modernism. Opponents, by contrast, favour radical modifications or demolition to make way for large-scale property redevelopment. Within this debate a very important aspect goes missing, namely what social purpose this space ought to fulfil. Our approach to this debate is to focus on the *modernist* aspect of these buildings, and contextualize the architectural and planning concepts they represent within the broader philosophical ideas of the modernist movement. What does it mean to "preserve" modernist architecture?

In Germany the prevailing approach to heritage preservation (*Denkmalpflege*) by public authorities relates to listed buildings and monuments – that is, physical structures that are officially designated as "testimonies to the past" (Oevermann and Mieg 2014b, 5). This contrasts with other conceptions of heritage that stress cultural practices related to a particular site or the community of stakeholders as the actor

defining what heritage is (Oevermann and Mieg 2014a, 15). As a production of space, preservation and conservation efforts thus involve
several dimensions that can be distinguished as what Henri Lefebvre
(1991) referred to as perceived, conceived, and lived space. These differences also account for conflicts among actors who have diverging
intentions around what heritage is supposed to be.

Modernism: Broad and Narrow Conceptions

In urban planning and architecture, the term "modernism" often refers
to a rather narrow and specific understanding of the concept. It is
important, however, to keep a broader notion of modernism in mind as
an intellectual backdrop; it poses a challenge that the narrower version
has been just one historical attempt at addressing.

Our broader conceptualization of modernism draws heavily from
Marshall Berman. "Modernism," according to Berman, refers to the set
of visions and approaches responding to a "mode of vital experience ...
that is shared by men and women all over the world today" (Berman
1982, 15). Modernism refers to human attempts to shape the future,
without denying the continual maelstrom of modernity that subjects
purposes, methods, and meanings to an ongoing process of change. "To
be modern," according to Berman, "is to find ourselves in an environment that promises us adventure, power, joy, growth, transformation
of ourselves and the world – and, at the same time, that threatens to
destroy everything we have, everything we know, everything we are"
(15). Berman's conception of modernism thus is paradoxical: The osten-
sible aim to assert dignity and find oneself at home in modern times is
consistently undermined by the modernizing process in which (following Marx's dictum) "all that is solid melts into air." The challenge then
is, in Baudelaire's words, to "extract the eternal from the ephemeral"
(Baudelaire 1986, 37).

Berman thus makes a useful distinction between modernism, modernization, and modernity. These moments are distinct yet in a dialectical relationship to each other. Modernism seeks to guide the process of
modernization and bring these multifaceted developments into some
kind of coherence with a specific purpose. At the same time, the process
of modernization continually changes the conditions for conceiving of
a modernist project. The experience of this dynamic is called "modernity," and in turn shapes how subjects engage in these modernizing
processes as well as their (modernist) ideas about them.

In contrast to this philosophical conception, much of the scholarly
writing on urbanism and urbanization from the 1960s on confronts a

more narrowly conceived notion of modernism, especially the undemocratic imposition of a particular social vision on urban citizens (Jacobs
1961; Sandercock 1998a, 1998b) that is sometimes referred to as "high
modernism" (Harvey 1990). This kind of modernism became politically hegemonic in the first three postwar decades in Western capitalist
and Soviet socialist countries as well as beyond. The practice of high
modernism was clearly facilitated by the postwar context in which
European cities destroyed in war had to be urgently reconstructed
while at the same time these reconstruction efforts faced severely limited resources, thus calling for efficient and no-frills solutions. The
destruction also functioned as a pretext for a slash-and-burn approach
that arguably sacrificed certain areas that might have been rehabilitated
for the sake of implementing visions of urban functionalism, as para-
digmatically espoused by the Athens Charter.[1] Particularly in war-torn
societies, such an approach was also justified by the high modernist
desire to overcome the past and the belief in the promise of a freer and
more exciting future that would "wipe away whatever came earlier in
the hope of reaching at last a true present ... a new departure" (Paul
de Man, cited in Berman 1982, 332).[2] This high modern vision distinguished four essential urban functions (dwelling, recreation, work,
and transportation) that were to be spatially separated to improve the
quality of everyday life. Cities were conceived of and built as "living
machines" (Harvey 1990, 31) to be modelled rationally in analogy to the
"factory" of Fordism.

Alexanderplatz: The Modernism of a Socialist Exemplar

Urban redevelopment in socialist countries like the GDR shows many
similarities with high modernist visions, but it also has some distinctive
features. Alexanderplatz offers insights into the GDR's complicated
relationship with modernism: Modernist ideas are clearly featured in
the GDR design of Alexanderplatz, yet in a particular form. Officially,
the GDR regime disavowed modernism as a bourgeois cultural phenomenon. Modernism offered a foil, often in the form of a strawman,
against which the regime efforts of building a socialist city were con-
trasted as if they were an entirely different endeavour. At the same time,
the goal of building the socialist city shared many ambitions with modernist projects; it aimed at resolving the negative social consequences of
the "creative destruction" of the postwar city – alienation and economic
misery – and through collectivization and a rational ordering of the
economy, it embraced the modern promise of growth, development,
and a better society.

The plans for Alexanderplatz were presented by the architects' collective of Peter Schweizer, Dorothea Tscheschner, Dieter Schulze, and Erwin Schulz in 1964 and displayed clear parallels to international developments in modernism. Tscheschner herself considers Alexanderplatz the "architectural high point" of modernism in the GDR (Tscheschner 2000, 268). Bodenschatz, Engstfeld, and Seifert (1995, 100) argue that since the late 1950s the course for architecture and urban development corresponds to a general model that both East and West were oriented towards. Kip, Young, and Drummond (2015) analyse the specific confluence of modernist and socialist ideas that shaped the GDR's development of Alexanderplatz as a form of "socialist modernism." Officially, however, Alexanderplatz figured as a "socialist exemplar." Finished in 1971, Alexanderplatz was a key piece in the development of the government centre in Berlin (see figure 9.1).[3] Gisa Weszkalnys (2008, 253) notes that "socialist planning constructed the square as a model for other GDR cities and as an expression of a specific form of future socialist society." Claire Colomb (2007, 289) makes a similar assessment when she states that Alexanderplatz was "planned to symbolically display the spirit of socialism." Buildings at Alexanderplatz represented key elements for socialist progress (which at the same time could also be read as modern aspirations): The House of Electrical Industry, The House of the Teacher, The House of Travel, The House of Statistics, and the iconic TV tower embodying telecommunications. As a "square of the people," "Alex" was also to cater to modern cravings for consumption, particularly through the largest department store in the GDR (Centrum Warenhaus) with the most refined assortment of consumer goods. As a central traffic hub, Alexanderplatz was committed to the modernist emphasis on facilitating traffic flow. Besides various modes of public transportation, including trains, subways, trams, and buses, that connected at Alexanderplatz, the three adjacent major boulevards signalled a commitment to car traffic. Just outside of these buildings, several large (eight- to eleven-storey) housing estates were built to present the regime's high aspirations for socialist living standards for the population. In the 1960s, the surrounding housing developments, produced using industrial methods, counted internationally among the technologically most advanced mass housing complexes at the time (Leinauer 2004, 121).

The largesse of the 1960s Alexanderplatz redevelopment ostensibly shares with high modernist planning visions the idea of building the city of the future on a large scale – and from scratch. In fact, the enormous scope of the redevelopment plan may be considered a "hyper-modernist" approach that dwarfed many modernist planning

Bundesarchiv, Bild 183-H1002-0001-016
Foto: Sturm, Horst | 2. Oktober 1969

9.1 Alexanderplatz, 1969. Source: Bundesarchiv, Bild 183-H1002-0001-016;
Photographer: Horst Sturm, 2 October 1969.

efforts in the West. In comparison to the pre-war experience of overcrowded, unsanitary living conditions in Berlin that were still present among Berliners in the 1960s, it is likely that such spaciousness
appeared like a coup de libération.

Moreover, the socialist centrality of Alexanderplatz departed from
the ideas of CIAM as expressed in the Athens Charter. The redevelopment followed the GDR's "Sixteen Principles of Urban Development"
that, "in contrast to the 'Athens Charter,'" as Tscheschner (2000, 260)
notes, "took on traditional ideas of a compact city with closed streets
and squares as well as a centre with dominant buildings as the starting
point for urban planning." Alexanderplatz was to become a key embodiment of this planning vision. The adjacent TV tower, at 368 metres
the tallest structure in Germany and one of the tallest in the world at
the time, together with the high-rise hotel on the square, underlined the
importance (centrality) of Alexanderplatz.

After *Wende* – The Debates in the 1990s

In German, *Wende* means "turning point," in this case the collapse of the socialist GDR. The devaluation of the buildings at Alexanderplatz as well as other modernist buildings in the former GDR after the Wende can be related to three larger processes – cultural, political, and economic – that intersected in a mutually reinforcing way.

On the cultural level, prior to the Wende modernist architecture and buildings in the West had already lost professional and public appeal and had largely fallen into disgrace (for an account of the complex reasons, see Harvey 1990). In 1987, an international building exhibition in West Berlin celebrated the idea of "careful urban renewal" to restore late nineteenth-century buildings for aesthetic, economic, and social reasons, rather than replacing them with modernist buildings. In the 1980s, even the GDR experimented with such restorative approaches and neo-historicist reconstruction of buildings, such as in the Nikolaiviertel, the historic core of the city (Urban 2007). In West Berlin, these explorations were solidified and theorized further under the label of "critical reconstruction," a term coined by architect Josef Paul Kleihues. Despite the conceptual ambiguity of critical reconstruction (Murray 2008), or maybe because of it, its ideas gained hegemonic status within the political and planning elite, particularly in the 1990s. The pretence of this approach was to reconstruct a "Berlinian architecture" as well as to emphasize Berlin's character as a "European city," although that character remained vague.[4] This historicist look back was highly selective in idealizing the pre-First World War Wilhelminian era and carefully overlooking the modernism of 1920s Weimar Berlin. The post-Second World War modernist architecture was heavily criticized for its lack of concern for history and tradition and its realization in many cases through slash-and-burn redevelopment. Former building senator of Berlin, Hans Stimmann, summarizes the motivation for "critical reconstruction" as follows: "I think that is one of the points ... that this generation that's living so much in the virtual world ... needs all the more real places. At the very latest, by the time they have children they begin to search: Where are the roots? Where is the city really? I am quite sure that there is a need for a common centre."[5]

Paul Sigel notes that the discussion and the devaluation of modernist buildings also affected former spearhead projects in West Berlin (including the Hansaviertel, the Kulturforum, and Ernst-Reuter Platz), but "ultimately, a critical engagement with late modernist structures focused on the eastern part of the city, in which, on the one side, the need for restoration was greatest, and, on the other side, the encompassing need for investment required fundamental strategic decisions with the existing substance" (Sigel 2009, 106).

In this respect then, anti-modernist sentiments went hand-in-hand at the political level with the endeavour to establish a new hegemony in the territory of the former GDR. Bruno Flierl describes the corresponding urban policy strategy adopted in East Berlin and the other new states of East Germany like this: "the aim was to purge the East German past in practical and aesthetic terms from the newly appropriated cities. At the same time, the goal was to create, in each of the New States with its old cities, architectural signs that would symbolize the new leading political and economic institutions and impress upon the inhabitants of every town that they were now living in a new state: the Federal Republic of Germany" (Flierl and Marcuse 2009, 269–70).

At the economic level, the new political reality was only too eagerly exploited by investors seeking to make windfall profits from the expansion of the land and real estate market. The cultural and political devaluation of the former territory and its physical structures opened the prospect for cheap purchases, state subsidies, and the possibility to completely redo the terrain, particularly in central hotspots of the city. In the early 1990s, a feverish run on land and real estate took place as demographic growth projections predicted that the number of inhabitants would almost double from 3.5 million to 6 million within two generations (Lehrer 2006, 333). Some boosters presaged Berlin to become a global city, with global connections competing to make investments in this strategic site connecting Western Europe with the new markets of Eastern Europe.

The confluence of all three processes had significant effects on the post-Wende debates around redeveloping Alexanderplatz. In the first few years after reunification, the square was left to deteriorate and "became an apt vehicle for talking about the demise of the GDR state and the future it once embodied" (Weszkalnys 2007, 211). Tscheschner explained in an interview that during the GDR approximately 100,000 employees worked in the area and that after reunification they were let go in a matter of months.[6] Shops and restaurants at Alexanderplatz were gradually closing down, thus fostering the impression of Alexanderplatz as an "empty" space that was decaying physically and, with the increasing presence of punks and other street folk, presumably socially as well. The areas around the square at Alexanderplatz were considered for a while to be the largest outdoor urinal and the largest outdoor garbage receptacle in all of Berlin! The vast open spaces, framed by the large areas of the square as well as the broad streets, were soon likened by developers and media commentators to "Siberian Plains" (Weszkalnys 2010). The square and the building composition itself were criticized as an example of "socialist monumentality and strategy to overwhelm [*Überwältigungsstrategie*]" (Sigel 2009, 105).

This seeming "waste of space" at Alexanderplatz in terms of physical and social dimensions and its deterioration made the situation seem urgent and requiring redress. In this context of fantastic expectations of growth, investors presented themselves as coming to the rescue: they would bring this central location and traffic hub to a new renaissance as a key site in Berlin's global city strategy. The enormous state-owned space appeared like an enormous opportunity for investors and politicians to develop at a large scale and without many obstacles. Increasingly, the significance of Alexanderplatz was also pushed politically. Former senator for urban development Peter Strieder underscored the political significance in the strongest of terms, stating that "Alexanderplatz, as a project in the centre of Berlin, will be a special symbol of the inner reunification so important for the identity of Berliners" (Strieder 2001, 3).

In 1992 an architectural competition for the redevelopment of Alexanderplatz was announced. In this competition, only one building composition from the socialist period was considered as heritage and thus had to be left in the plan (the House of the Teacher cum Congress Hall), thus opening enormous possibilities for investment.[7] Reflecting the decisive influence of investors within the competition's jury, the entries outdid each other in the amount of office and retail spaces proposed. One such exaggerated entry by architects Hans Kollhoff and Helga Timmermann was selected in 1993 as the winning proposal. Their original scheme calling for thirteen high-rises (each 150 metres in height) was later reduced to ten, but even in this modified form it far outpaced in height, size, and investment volume the plans for the other two central city locations of Berlin that were deemed suitable for tall buildings, Breitscheidplatz and Potsdamer Platz. The Kolhoff scheme, with 1.3 million square metres of floor space for office, retail, and some residential, amounted to a huge projected investment (Senatsverwaltung für Stadtentwicklung und Wohnen Berlin 2017a). It therefore seemed that Alexanderplatz was a particularly fierce case in which the post-reunification derision of "socialism" led to the space becoming designated as a *tabula rasa* (see figure 9.2).

Although experts had earlier criticized the post-reunification investment and redevelopment euphoria as unfounded, by the mid-1990s it became evident even to the biggest boosters that such expectations were grossly exaggerated. Germany entered into a recession and the building economy that had driven Berlin's economy was cooling off significantly. The outlook for the realization of the Kollhoff proposal

9.2 Model of Alexanderplatz district with the Kolhoff proposal for ten high-rise towers just to the right of the centre. Photo: Douglas Young.

became increasingly dim and thus far none of the high-rises has been built. However, even when the economic prospects and the demographic projections for Berlin were readjusted downwards, the Berlin Senate held on to the Kollhoff plan.[8]

Current Debate

To this day, more than two decades after the adoption of the Kolhoff Plan for Alexanderplatz, only a moderate physical modification of the area has actually been undertaken. However, business has clearly improved at Alexanderplatz in recent years with the renovation (and extension) of the former Centrum Warenhaus (completed in 2006), the new Alexa shopping mall across the street from the square (2007), the construction of a new building for an electronics store on the north-eastern corner of the square (2009), and the department store building Alea 101 on the other side of the train station (2014).

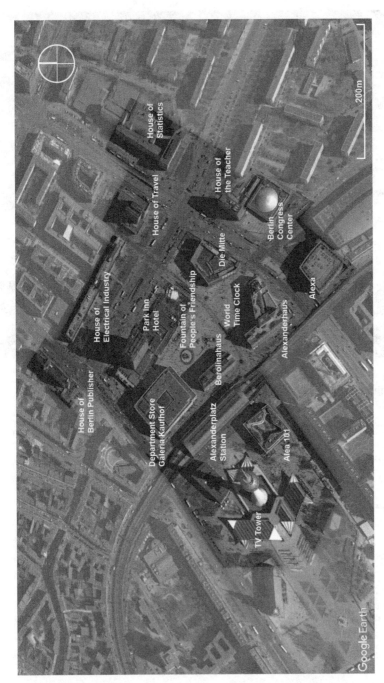

9.3 Alexanderplatz satellite image with key buildings marked. Source: Katharina Frieling.

Alexanderplatz today in no way resembles the "little Manhattan" of the Kollhoff proposal. In spring 2013, one of the investors announced that they would go forward with building one residential high-rise to address the growing demand for luxury housing in the city. The architectural proposal for this building by "starchitect" Frank Gehry was officially selected in January 2014 (Paul 2014). In 2016, another residential high-rise, called Capital Tower, with ten floors of furnished and "serviced apartments," was announced. However, voices against such developments have also grown louder. Only about a month after the plans for the first condo were announced, the building senator of Berlin, Regula Lüscher, suggested changing the development plan as a way of coming to terms with the unlikelihood of investment interest to realize the high-rises. She argued for integrating the existing GDR buildings and criticized the Kollhoff plans: "The GDR plan with its high-rises that related to each other was sidelined by the scale of the 150-metre-tall buildings. Now is a good time to make a plan that is more strongly oriented towards the already existing buildings" (Paul 2013b). Lüscher's controversial statements have been debated by planners, architects, developers, and politicians, and stirred a considerable discussion within the press. What is interesting about this debate is that the critique of modernism is not at the centre – instead questions of historic memorialization and economic development are. At first glance, the debate seems to be polarized, and journalist Ralf Schönball (2013a) claims that it "divides the architectural community" and beyond.

On the one side, Lüscher's supporters point to – and demand – a growing appreciation of the modernist heritage of the GDR. Prominent among them is the president of the Berlin Architects Association (Architektenkammer Berlin), Christine Edmaier (Loy 2013).[9] Advocates for preservation concentrate their efforts particularly on the House of Travel. The former vice-president of the Berlin Architects Association, Theresa Keilhacker adds that, in particular, the House of Travel with its "sculptural architecture and its unique projecting roof definitely deserves to be landmarked" (Schönball 2013b). The same newspaper report also cites "experts" who emphasize the building's significance as an authentic GDR structure.

Between 2013 and 2015, the heritage preservation agency of the State of Berlin reviewed the remaining structures at Alexanderplatz built during the time of the GDR. According to news reports, already in 2013 the head of the agency, Jörg Haspel, attested that there was a good chance especially for the House of Travel, because its "ideological and artistic content" can be read particularly easily (Schönball 2013b). The frieze on its base by Walter Womacka, entitled "Humanity Overcomes Time and Space," is an invitation to frontier-crossing travels (ironically, since for most GDR citizens the GDR border just a kilometre away already meant an end to movement). It corresponds to the wall mosaic at the House of the Teacher that was also conceived by Womacka. The suggestion that the House of Travel and the House of the Teacher appear to create a gate at a central traffic hub (and at the site of a former town gate) further underscores the argument that since the House of the Teacher is already protected, it is important to preserve the House of Travel as well. Haspel further distinguishes it as an ambitious example of GDR "mass architecture" (Schönball 2013b). Architect and professor Wolf Eisentraut (2013) applauds the initiative of heritage preservation and claims that it is not about "a one-sided aesthetic judgement and even less about taste, but the conservation of built testimonies of cultural identity of completed epochs."

On the other side, those who mobilize against such preservation attempts take on the rhetoric of modernization. Architect Jan Kleihues, who has spoken out in favour of holding on to the Kollhoff plan, underscores his demand by pointing to the negative consequences of the 220-metre-long facade of the House of Electrical Industry. By blocking access between Alexanderplatz and the area to the north, the building effectively undermines urban and economic development (Schönball 2013c).

Other present-day modernizers emphasize aesthetic aspects. Journalist Peter von Becker (2013) complains that conservation efforts "mistake the challenges of vital urban development for the archives of a historical museum." In an op-ed piece, von Becker polemicizes that the desire to preserve GDR architecture is an effort to keep Berlin similar to Pyongyang. Without making this comparison concrete, he invokes stereotypical views of Pyongyang as a politically oppressive and economically backward environment. To avoid such a fate for Berlin, von Becker proposes that "it is required to awaken a fundamental sense of beauty among Berlin's politicians, heritage preservationists, and urban developers" (n.p.). This view of socialism as anti-modern is also expressed in the op-ed piece by another architectural critic, Danckwart Guratzsch (2013), in the national newspaper *Die Welt*. Here, Alexanderplatz is presented as an instance of the "fallow lands and inner-city architectural abberations [*Bausünden*] that socialism has left behind." Considering the GDR buildings at Alexanderplatz to be "miserable" (apparently excepting Henselmann's House of the Teacher) another opinion-maker of *Die Welt*, Rainer Haubrich (2013) considers them useful "at the most to confront Berlin visitors with the absurd dimensions and the scary character of socialist city building – not to speak of the abysmal details of these buildings." Clearly, the polemical acerbity of these commentators is reminiscent of the architectural debates in first few years following reunification.

Finally, in 2015, the heritage preservation agency of Berlin listed the House of Travel and the House of Berlin Publisher together with the World Time Clock on Alexanderplatz. As building senator Lüscher declared, the intention was to showcase the "numerous qualities" of architecture built in East Berlin during the Cold War. As the ideological opposite to West Berlin's Breitscheidplatz and Ernst-Reuter-Platz, squares that had already been placed under heritage protection, the newly listed buildings of East Berlin's Alexanderplatz are a testimony to the "double Berlin" of the time. Today, however, the municipal government continues to hold on to the masterplan, albeit now with only nine high-rises (Senatsverwaltung für Stadtentwicklung und Wohnen Berlin 2017b).

Conclusion: Modernity against Modernism?

A closer look at the debate reveals a more differentiated picture than a simple division between aesthetic concerns for historic preservation and economic concerns for redevelopment. Contemporary positions in the political arena rarely argue for preservation or redevelopment *in toto* and instead focus on the merit of individual buildings. A long-time advocate for the preservation of GDR buildings, Thomas Flierl, argued for the preservation of buildings such as the House of Travel or the House of Berlin Publisher, but also accepts the necessity of making physical alterations in the House of Electrical Industry to make a connection with the housing settlement north of Alexanderplatz and facilitate urban development (Schönball 2013b). Some commentators thus prefer preservation of selected buildings in conjunction with the development of some high-rises at Alexanderplatz, such as condos (Maak 2013). The obvious problem with an approach focusing on individual buildings, however, is that it doesn't do justice to their original conception as part of a larger assemblage. And as Wolf Eisentraut (2013) contends, the assemblage has already been significantly altered.

Moreover, preserving modernist buildings as landmarks further seems to be quite at odds with the high modernist aspirations for ongoing development. Thus, in this debate about who "the moderns" are, the tables seem to be turned. Those who argue against landmark status for such GDR modernism advance ideas of vitality, urbanity, and development for Alexanderplatz – tropes that come from the modernist vocabulary. By contrast, those who advocate for landmark preservation seem to concentrate on purely formal aesthetic aspects of "high" modernism. Nikolaus Bernau (2013) proposes a more modernist approach to historic preservation that relates to "lived experiences." Responding to von Becker's Pyongyang polemic, Bernau makes the plea that "historic conservation must be displeasing ... Every generation ought to be able to add its lived experience to the landmark lists – at the risk of being insulted by those who have forgotten the landmark-fights of their youth, as defendants of Pyongyang. It isn't Pyongyang standing at Alexanderplatz, but the real, lively, historical Berlin." For Bernau landmark preservation is motivated by a desire to find a feeling of security, to bring back the past in a tumultuous present. However, in our view, such intentions would not do justice to the modernist spirit either.

Considering heritage as a produced space in a Lefebvrian sense, we want to bring up the question of the third dimension of space, the space's social purpose. What are the visions and promises of modernist spaces, and what does it mean to preserve them? In other words, what would it mean to realize the promise of Alexanderplatz as designed during the GDR? This question has been sidelined by a specialist discourse on aesthetic standards, economic dynamics, and a politics of museumization. Alternative social visions are absent in the current debate. On the one side, advocates for the "modernization" of Alexanderplatz make only superficial gestures towards democracy or equality. An investor who is currently in the process of developing the first high-rise based on the Kollhoff plan quite frankly claimed to be aiming primarily at developing housing for the upper-market sector. "Social housing is not possible in a high-rise," a representative stated at a discussion forum (Paul 2013a). On the other side, advocates for preservation also show few concerns for social issues or the underlying visions behind the buildings. So far, there are no particular ideas for how these buildings that are to be landmarked should be used – beyond their current usage by "creative" enterprises like architectural offices, co-working spaces, or a night club (Gutzmer 2013). What is absent from this debate are in-depth discussions with the current "users" of Alexanderplatz, and with those who might be excluded from the space. What are the experiences and hopes attached to this space for the many thousands of workers, tourists, shoppers, commuters, residents, punks, and skateboarding youth who are present in Alex every day? Thinking about how these different views may be negotiated and, at least partially, reconciled in a wide-ranging vision seems to do more justice to the modernist spirit – whether it preserves the physical appearance of Alexanderplatz or not.

NOTES

1 CIAM, The Congrès Internationaux d'Architecture Moderne (International Congresses on Modern Architecture), was an organization of modernist architects established in 1928 and disbanded in 1959. Every few years CIAM held a congress to discuss issues of architecture and urban planning. CIAM IV, held in 1933, discussed the theme of the functional city, the concepts of which were later published by Le Corbusier as the Athens Charter. The influence of the Athens Charter was felt around the world and lasted for several decades.

2 Berman cautions the reader not to take the disregard for the past and the effort to wipe it out as a characteristic of modernism per se. Many modernists, like Benjamin, have emphasized the significance of history as an accumulation of catastrophes that weighs on the present. A liberating engagement with the present means coming to terms with the unfulfilled promises of the past that still weigh on the present. Considering Jane Jacobs and the movement against slash-and-burn urbanism as a modernist movement, Berman (1982, 318) argues that "Jacobs' point is that the so-called modern movement has inspired billions of dollars' worth of 'urban renewal' whose paradoxical result has been to destroy the only kind of environment in which modern values can be realized. The practical corollary of all this – which sounds paradoxical at first, but in fact makes perfect sense – is that in our city life, for the sake of the modern we must preserve the old and resist the new."

3 The development of Alexanderplatz was the culmination of a long and contorted process of planning the central areas of the city. The heavy war destruction of inner-city areas had facilitated large-scale planning for a new centre of the socialist state that was to extend roughly from Brandenburg Gate along Unter den Linden across the River Spree into Alexanderplatz.

4 The idea of the "European city," according to Sigel (2009, 107 our translation) "if concretized in planning discussions at all, was associated with a compact building of the centre, a clear division between public and private spaces, mixed-use as well as with limited building heights." Henceforth, when quoting a German source, the translation is ours.

5 Interview with Hans Stimmann, 13 July 2011, Berlin.

6 Interview with Dorothea Tscheschner, 13 July 2011, Berlin.

7 The two Behrens buildings from the 1930s were also designated as heritage.

8 The investors at Alexanderplatz were glad when the government continued to push forward with the passing into law of the megalomaniac development plan, even as the economic outlook became dark. Without having made any significant investment, the finalization of the planning process in

2000, at least for the largest part of the area, brought the investors an estimated tripling of the land's value (Lenhart 2001, 15).

9   An example of such growing interest in Eastern modernist architecture
    ["*Ostmoderne*"], especially among young architects and artists, is the work
    by Butter and Hartung (2004).

REFERENCES

Baudelaire, Charles. 1986. *My Heart Laid Bare and Other Prose Writings*.
    Translated by Norman Cameron, edited by Peter Quennell. London, UK:
    Soho Book Co.

Berman, Marshall. 1982. *All That Is Solid Melts into Air: The Experience of
    Modernity*. New York, NY: Simon and Schuster.

Bernau, Nikolaus. 2013. "Architektur in Berlin: Ein Denkmal ist mehr als das,
    was gefällt." *Berliner Zeitung*, 16 August. http://www.berliner-zeitung
    .de/berlin/architektur-in-berlin-ein-denkmal-ist-mehr-als-das--was
    -gefaellt,10809148,24031854.html

Bodenschatz, Harald, Hans-Joachim Engstfeld, and Carsten Seifert. 1995.
    Berlin auf der Suche nach dem verlorenen Zentrum. Hamburg, Germany:
    Junius.

Butter, Andreas, and Ulrich Hartung, eds. 2004. *Ostmoderne: Architektur in
    Berlin 1945–1965*. Berlin, Germany: Jovis.

Colomb, Claire. 2007. "Requiem for a Post Palast: 'Revanchist Urban Planning'
    and 'Burdened Landscapes' of the German Democratic Republic in the New
    Berlin." *Planning Perspectives* 22, no. 3: 283–323. https://doi.org/10.1080
    /02665430701379118

Eisentraut, Wolf. 2013. "Alexanderplatz: Das Scharnier im Osten." *Der
    Tagesspiegel*, 25 September. http://www.tagesspiegel.de/kultur
    /alexanderplatz-das-scharnier-im-osten-berlins/8847284.html

Flierl, Bruno, and Peter Marcuse. 2009. "Urban Policy and Architecture for
    People, Not for Power." *City* 13, no. 2–3: 264–77. https://doi.org/10.1080
    /13604810902982235

Guratzsch, Dankwart. 2013. "Baut zehn Wolkenkratzer an den Alexanderplatz!"
    Welt Online, 14 April. http://www.welt.de/kultur/kunst-und-architektur
    /article115281061/Baut-zehn-Wolkenkratzer-an-den-Alexanderplatz.html

Gutzmer, Alexander. 2013. "Denkmalschutz-Debatte in Berlin: Rettet den
    Alexanderplatz! Rettet die Ostmoderne!" *FOCUS Online*, 29 August. http://
    www.focus.de/politik/experten/gutzmer/denkmalschutz-debatte-in
    -berlin-rettet-den-alexanderplatz-rettet-die-ostmoderne_id_3125953.html

Harvey, David. 1990. *The Condition of Postmodernity: An Enquiry into the Origins
    of Cultural Change*. Oxford, UK: Blackwell.

Haubrich, Rainer. 2013. "Kein Denkmalschutz für die elenden DDR-Bauten!" *Welt Online*, 13 August. http://www.welt.de/kultur/kunst-und-architektur /article118975328/Kein-Denkmalschutz-fuer-die-elenden-DDR-Bauten .html#disqus_thread

Jacobs, Jane. 1961. *The Death and Life of Great American Cities*. New York, NY: Vintage.

Kip, Markus, Douglas Young, and Lisa B. Welch Drummond. 2015. "Socialist Modernism at Alexanderplatz." *Europa Regional* 22, no. 1–2: 13–26.

Lefebvre, Henri. 1991. *The Production of Space*. Oxford, UK: Blackwell.

Lehrer, Ute. 2006. "Willing the Global City: Berlin's Cultural Strategies of Interurban Competition after 1989." In *The Global Cities Reader*, edited by Neil Brenner and Roger Keil, 332–41. New York, NY: Routledge.

Leinauer, Irma. 2004. "Das Wohngebiet Karl-Marx-Allee: Industrielles Bauen zwischen Strausberger Platz und Alexanderplatz." In *Ostmoderne: Architektur in Berlin 1945–1965*, edited by Andreas Butter and Ulrich Hartung, 114–23. Berlin, Germany: Jovis.

Lenhart, Karin. 2001. *Berliner Metropoly: Stadtentwicklungspolitik im Berliner Bezirk Mitte nach der Wende*. Berlin, Germany: Springer Verlag für Sozialwissenschaften.

Loy, Thomas. 2013. "Bebauung am Alexanderplatz in Berlin-Mitte: Präsidentin der Architektenkammer fordert neuen Masterplan." *Der Tagesspiegel*, 31 May. http://www.tagesspiegel.de/berlin/bebauung-am-alexanderplatz -in-berlin-mitte-praesidentin-der-architektenkammer-fordert-neuen -masterplan/8278414.html

Maak, Niklas. 2013. "Berlin Alexanderplatz Die Wahrheit der Türme." *Frankfurter Allgemeine Zeitung*, 14 April. http://www.faz.net/aktuell /feuilleton/berlin-alexanderplatz-die-wahrheit-der-tuerme-12147588.html

Murray, George J.A. 2008. "City Building and the Rhetoric of 'Readability': Architectural Debates in the New Berlin." *City and Community* 7, no. 1: 3–21. https://doi.org/10.1111/j.1540-6040.2007.00238.x

Oevermann, Heike, and Harald A. Mieg. 2014a. "Studying Transformations of Industrial Heritage Sites: Synchronic Discourse Analysis of Heritage Conservation, Urban Development, and Architectural Production." In *Industrial Heritage Sites in Transformation: Clash of Discourses*, edited by Heike Oevermann and Harald A. Mieg, 12–25. New York, NY: Taylor and Francis.

– 2014b. "Transformations of Industrial Heritage Sites: Heritage and Planning." In *Industrial Heritage Sites in Transformation: Clash of Discourses*, edited by Heike Oevermann and Harald A. Mieg, 3–11. New York, NY: Taylor and Francis.

Paul, Ulrich. 2013a. "Diskussion über Zukunft des Alexanderplatzes." *Berliner Zeitung*, 29 May. http://www.berliner-zeitung.de/alexanderplatz

/leserforum-berliner-zeitung-diskussion-ueber-zukunft-des-alexanderplatzes
,22289878,22901710.html

– 2013b. "Debatte um den Alexanderplatz: 'Im Grunde ist der Plan nicht
umsetzbar.'" *Berliner Zeitung*, 4 November. http://www.berliner-zeitung
.de/alexanderplatz/debatte-um-den-alexanderplatz--im-grunde-ist-der
-plan-nicht-umsetzbar-,22289878,22334836.html

– 2014. "Entwurf von Frank Gehry gewinnt: Größtes Hochhaus Berlins ent-
steht am Alexanderplatz." *Berliner Zeitung*, 27 January. http://www
.berliner-zeitung.de/berlin/entwurf-von-frank-gehry-gewinnt-groesstes
-hochhaus-berlins-entsteht-am-alexanderplatz,10809148,26006380.html

Sandercock, Leonie. 1998a. "The Death of Modernist Planning: Radical Praxis
for a Postmodern Age." In *Cities for Citizens: Planning and the Rise of Civil
Society in a Global Age*, edited by Mike Douglass and John Friedmann,
163–84. Chichester, UK: John Wiley and Sons.

– 1998b. *Towards Cosmopolis: Planning for Multicultural Cities*. New York, NY:
J. Wiley.

Schönball, Ralf. 2013a. "Neugestaltung des Alexanderplatzes: Pläne für den
Alex spalten Berliner Architektenschaft." *Der Tagesspiegel*, 4 April. http://
www.tagesspiegel.de/berlin/neugestaltung-des-alexanderplatzes
-plaene-fuer-den-alex-spalten-berliner-architektenschaft/8021582.html

– 2013b. "Berlin will DDR-Bauten am Alex unter Denkmalschutz stellen." *Der
Tagesspiegel*, 13 August. http://www.tagesspiegel.de/berlin/staedtebau
-berlin-will-ddr-bauten-am-alex-unter-denkmalschutz-stellen/8629648.html

– 2013c. "Denkmalschutz für TLG-Bauten? Alexanderplatz: Der Masterplan
wackelt." *Der Tagesspiegel*, 15 August. http://www.tagesspiegel.de/berlin
/denkmalschutz-fuer-tlg-bauten-alexanderplatz-der-masterplan
-wackelt/8640672.html

Senatsverwaltung für Stadtentwicklung und Wohnen Berlin. 2017a.
"Alexanderplatz." http://www.stadtentwicklung.berlin.de/planen
/staedtebau-projekte/alexanderplatz/

– 2017b. "Denkmalschutz für die Nachkriegsmoderne am Alexanderplatz"
[press release]. http://www.stadtentwicklung.berlin.de/aktuell
/pressebox/archiv_volltext.shtml?arch_1507/nachricht5680.html

Sigel, Paul. 2009. "Ein Platz. Viele Plätze. Projektionen und Spurensuche am
Berliner Alexanderplatz." In *Imaginationen des Urbanen: Konzeption, Reflexion
und Fiktion von Stadt in Mittel-und Osteuropa*, edited by Arnold Bartetzky,
Marina Dmitrieva, and Alfrun Kliems, 81–118. Berlin, Germany: Lukas Vlg
f. Kunst-u. Geistesgeschichte.

Strieder, Peter. 2001. "Alexanderplatz Berlin." In *Alexanderplatz Berlin:
Geschichte, Planung, Projekte – History Planning Projects*, edited by Annegret
Burg and Senatsverwaltung für Stadtentwicklung Berlin, 3. Berlin,
Germany: Kulturbuch-Verl.

Tscheschner, Dorothea. 2000. "Sixteen Principles of Urban Design and the Athens Charter?" In *City of Architecture of the City: Berlin 1900–2000*, edited by Thorsten Scheer, Josef Paul Kleihues, and Paul Kahlfeldt, 259–70. Berlin, Germany: Nicolai.

Urban, Florian. 2007. "Designing the Past in East Berlin before and after the German Reunification." *Progress in Planning* 68, no. 1: 1–55. https://doi .org/10.1016/j.progress.2007.07.001

von Becker, Peter. 2013. "Denkmalschutz in Berlin: Schrecken statt Schönheit." *Der Tagesspiegel*, 14 August. http://www.tagesspiegel.de/meinung /denkmalschutz-in-berlin-schrecken-statt-schoenheit/8635116.html

Weszkalnys, Gisa. 2007. "The Disintegration of a Socialist Exemplar Discourses on Urban Disorder in Alexanderplatz, Berlin." *Space and Culture* 10, no. 2: 207–30. https://doi.org/10.1177/1206331206298552

– 2008. "A Robust Square: Planning, Youth Work, and the Making of Public Space in Post-Unification Berlin." *City and Society* 20, no. 2: 251–74. https:// doi.org/10.1111/j.1548-744x.2008.00019.x

– 2010. *Berlin, Alexanderplatz: Transforming Place in a Unified Germany*. New York, NY: Berghahn Books.

PART THREE

Governance and Social Order

10 China's "New" Socialist City: From Red Aesthetics to Standard Urban Governance

CAROLYN CARTIER

Any traveller in China observing the urban landscape sees the symbolic signage of the "civilized city." Its forms include long red banners, emblazoned with outsized Chinese characters, draped across building facades or hung across streets as if celebrating a new building or special event. Civilized city signage also appears in the form of modern advertisements on subway concourses and billboards along expressways. Above entrances to train stations and other sites of public infrastructure, its repetitive reminders about civilized conduct appear with "welcome" announcements in scrolling displays of red light-emitting diodes. If we know how to read them (they never appear in English), we find that many of these messages promote a range of values and virtues with socialist characteristics. "Collectively strive to be civilized, together build a civilized city" (*Zhengzuo wenming ren, gongchuan wenming cheng* 争做文明人，共创文明城). The "collective" is a keyword in China's popular history that connects the fleeting rural past with the urbanizing present. Other messages appear more modern yet still code a city in the process of becoming. "Cultivate a civilized atmosphere, act as civilized citizens, and create a civilized city" (*Shu wenming xinfeng, zuo wenming shimin, chuang wenming chengshi* 树文明新风，做文明市民，创文明城市). No matter their style or context of appearance, dependably repetitive civilizing messages are ubiquitous reminders of the historic shift that has been taking place in China as hundreds of millions of people, negotiating unprecedented opportunities and inequalities, make new homes in new cities (figures 10.1 and 10.2).

Since the 1980s China has established over 450 new cities, yet the symbolic signage of the civilized city continues the style of Mao-era (1949–76) dissemination of Party messages. Typically referred to as "propaganda" (*xuanchuan* 宣传), such signage is usually disregarded in urban studies research on China. However, in the 2000s a special committee of the Chinese Communist Party (CCP) began formulating

10.1 "Strive to Become Civilized Citizens, Strive to Become a Civilized City." Shenzhen, 2005. Photograph by Carolyn Cartier.

10.2 "Establish a National Civilized City, Construct an International Metropolis." Xi'an, 2013. Photograph by Carolyn Cartier.

the "civilized city" concept for an applied governing project called the National Civilized City (*Quanguo wenming chengshi* 全国文明城市). Not a government program or a planning rationality in the conventional sense, the National Civilized City is an evolving policy regime based on Party thought work (Cartier 2013, 2016). It amalgamates contemporary social and economic reforms with the Chinese socialist tradition of didactic models (*dianxing* 典型 or *moshi* 模式) that represent exemplarity to govern conduct (Bakken 2000). It is a staunch reminder of the party-state's guiding role in the urban process in the People's Republic of China (PRC). Based on inter-textual assessment of Party documents, in this chapter I introduce the context, emergence, and scope of the National Civilized City. How does this socialist governing apparatus operate programmatically in contemporary China? The analysis follows the establishment of the program through the textual process of Party thought work, and its incorporation of modern urban standards and techniques of implementation and control. The chapter compares iterations of the National Civilized City Assessment System to see how the National Civilized City program entrains urban governance, through conduct of local officials, and incorporates Party leadership priorities, including priorities of Xi Jinping thought since 2012.

Locating the "Post" in China's Socialist Transformation

Apparent divergence between the "economy of appearances" (Tsing 2000) in China and Party programs like the National Civilized City asks for contextualization. The popular idea of Chinese cities serves up visual spectacles – novel architecture, megamalls, super-tall buildings, and forests of high-rise apartments towering over fading villages. In half a century China has replaced the Cultural Revolution's struggles over class politics with an urban revolution of state-contoured consumer capitalism. In reflection on the vast scale of transformation, the socialist-style policy programmatic of the civilized city would seem to be out of sync with the frenzied realization of China's globalizing condition. But it is this combination of capitalist-style cities with party-state administration of society and space whose conditions the National Civilized City program seeks to model, institutionalize, and govern.

In comparative perspective, the idea of China as post-socialist raises interesting questions because the PRC does not adopt a discourse of post-socialism. Post-1978 economic reform in China dismantled the top-down socialist planned economy while continuing top-down party-state authority. For instance, in 1982, at the Twelfth National Party Congress, Deng Xiaoping declared, "we must ... blaze a path of our own and build a socialism with Chinese characteristics" (Deng 1982). Since the PRC had taken the decision in December 1978 to open to the world economy, Deng's statement suggested elements of rote Party thought that might have thought to be history. But "socialism with Chinese characteristics" has become canonized in Party thought and widely repeated and republished (e.g. Lim 2014; Solé-Farràs 2008). In the Chinese political economy, economic reform has not been matched by political reform. The more recent turn of phrase, "capitalism with Chinese characteristics" (Huang 2008), recognizes China's post-socialist market economy that maintains strategic party-state authority including over state-owned enterprises, land ownership, and the establishment of new cities. Capitalism with Chinese characteristics points to the difference between marketization of the economy, with its production of capitalist forms in the built environment, and orthodox Party thought and its political capacities to spatially govern society.

When China's coastal cities opened to the world economy in the 1980s, the built environments of central business districts bore pre-war facades. Many were large industrial plants in urban cores. In the era of the socialist planned economy cities lacked commercial development thanks to Mao's infamous urban-economic dictum: "Turn consumer cities into producer cities" (*ba xiaofei chengshi biencheng shengchan chengshi* 把消费城市变成生产城市). Transforming the built environment of the Mao-era producer cities into new commercial landscapes required "demolition and relocation" (*chaiqian* 拆迁), including widespread removal of the socialist housing of many historic factory work units (Wilhelm 2004). Frequently, relocation has required removal to new high-rise development projects on distant ring roads. The process of restructuring restored commercial cores and built new suburbs – the key economic geography of the city as a post-socialist formation in China. This process makes the new commercial urban cores in historic cities emblematic of "urban transition" in China (Friedmann 2005; McGee et al. 2007; Tang and Chung 2002). The post-socialist economic city in China has emerged from the removal of historic industry and socialist housing to make way for urban services industries, "commodity housing," and the new consumer economy.

Through the turn of the century, general understandings of China's "reform and opening" largely treated China through the framework of economic transition, reflecting transformations in Eastern Europe after the fall of the Berlin Wall. The international scholarship scarcely considered how the CCP would maintain ruling power and revive socialist-era modes of policy and governance. In sync with economic predictions, the mainstream of international political economy continued to adopt assumptions about transition in China through the 1990s. But indicators of the problematic differences between economic and political conditions were at hand. In contrast to Europe, where "1989" is a metonym for a new era of openness, China's landmark 1989 event was the violent state suppression of public protest on 4 June in Beijing, whose history remains black-boxed or unexamined in China. While the Chinese economy globalized and Chinese cities realized consumer service centres, only inquiry into politics and political thought returned questions about how or in what ways the CCP would continue its socialist policies through the turn of the century.

Models and Model Cities

The National Civilized City is a model city policy program under the jurisdiction of the CCP's Spiritual Civilization Guidance Committee. But its conditions belie its titular connotations. It is not a general program or a program for or based on the capital, Beijing. Neither is it an award for design of the built environment or a conventional policy or planning model. The National Civilized City is the highest "honour" bestowed on a city in China. It is an award that demonstrates that a city has passed an external evaluation of standards, in multiple political, economic, social, and environmental categories, compiled by the Spiritual Civilization Guidance Committee. Its programmatic rationale yields a technology of governmentality that models uniform conditions through a set of dynamic goals, criteria of achievement, and indices of measurement. It is one of several "model city" programs in the contemporary PRC. Yet where other programs target improvement of particular conditions, the National Civilized City is an award for general improvement of the city's material and social conditions. Its agenda covers multiple arenas and categories of city administration, from "healthy and progressive human environment" and "solid and effective foundation" for policy implementation to "incorruptible and efficient environment" for public affairs. These aspirational conditions underscore how the National Civilized City honour is a different kind of award. Its texts, instructions, and operations seek to institutionalize, among Party and government officials, a process of not only striving to improve but socialization of a "moral order" in order to staunch the corruption that would limit China's transformation.

The idea of models and model conduct in China has a long history, and antedates the idea of model cities and the founding of the PRC in 1949. The idea of model behaviour in China traces back to the Confucian tradition of upholding learned men as exemplars of society

(Bakken 2000). Local histories of the imperial era routinely included sections featuring notable and exemplary personages at the county level, including a category for virtuous widows (Leung 1993). In the Mao era, models gained new traction through the Stakhanovite labour mobilization model, originally introduced to China from the Soviet Union in the 1940s. The image of Alexey Stakhanov, a mythologized model worker who symbolized socialist competition through zealously performed work, became transferred to collective labour in socialist China (Funari and Mees 2013). The labour model urged villages, factories, and industrial towns to collectively compete to increase production. Because exceeding quotas accrued rewards, masking failures and faking successes was common (Shapiro 2001).

Under reform, Party thought work has also promoted the idea of human "quality" (suzhi 素质) to urge "modernization" of rural populations under rapid urbanization (Anagnost 2004). China's best-known model citizen associated with moral quality is Lei Feng, a 1950s-era People's Liberation Army (PLA) soldier whose life continues to symbolize selfless labour contribution and unwavering socio-political dedication (Penny 2013). Historically, Lei Feng was "Chairman Mao's good soldier," whose loyalties defined the political sensibility of symbolic social relations between the people and the PLA and CCP. His model characteristics – unfailingly right-minded conduct and selfless acts – gained national and enduring proportions through hagiographical accounts (Larson 2009). That the myth and reality of his life continue to engender debate makes the model of Lei Feng more important in China's history than the man. In the reform era, since the 1980s, his symbolic capital has transformed. Images of Lei Feng now adorn various consumer goods, while his symbolic poses include volunteer activity dedicated to the betterment of urban society (Jeffreys 2012). Less a Party ideologue and now more an engaged urban citizen, the new Lei Feng – almost always wearing his PLA winter flaphat – cares about your quality of life (figure 10.3).

Maoist-era models and the political campaigns used to promote them constitute the historical and aesthetic roots of the contemporary model city programs. Such programs launched in 1997 with the China National Environmental Protection Model City. They include the National Entrepreneurial City, National Model City for Science and Technology, National Hygienic City, and National Excellent City for Comprehensive Management of Public Security (Zhao 2011). The current concept of the model city continues, in new ways, the use of model places in socialist planning. The two famous Mao-era model places were the Daqing oilfields and the Dazhai People's Commune. In the 1960s, the

10.3 "Lei Feng Memorial Day, March 5." Nanchang, Xinhua Bookstore, 2012. Photo copyright Vmenkov/Wikimedia, licensed under CC BY-SA 3.0. https://commons .wikimedia.org/wiki/File:nanchang_-_Bayi_Square_-_DSCF4203.JPG.

propaganda apparatus tirelessly urged the nation, "in industry, learn from Daqing" (*gongye xue Daqing* 工业学大庆), "in agriculture, learn from Dazhai" (*nongye xue Dazhai* 农业学大寨). Real and symbolic, these places were geographical contexts and constructs for reproducing a mode of governing that would socialize emulation and institutionalize repetition (Hoffman 2011, 57). Instead of basing a model on particular places, like Daqing and Dazhai, the contemporary National Civilized City program articulates a set of aspirational policies for all cities. What has not changed is that model city guidelines provide idealized visions of development and officially approved standards for modernization. Past and present, big red banners with CCP-approved political statements herald such model activities.

The first sign of the PRC's renewed interest in the ideological thought of "civilizing" emerged in the 1980s in response to new socio-economic problems in the early years of reform. In 1986 the Party Central Committee adopted the Resolution on Guiding Principles for Building a Socialist Society with Spiritual Civilization and launched an early "civilizing" campaign. Yet with international attention riveted to China's new international investment and foreign trade in coastal economic zones, the CCP campaign to control "spiritual civilization" seemed like an atavistic call by the Propaganda Department. But then in 1996, at the Fourteenth Party Congress, the CCP issued the Resolution Concerning Several Important Questions for Strengthening Socialist Spiritual Construction (CCP Central Committee 1996). The next year the Propaganda Department established the Central Guidance Committee for Building Spiritual Civilization, whose programs focus on socialization of state values for social control (Brady 2008). Where the spiritual civilization campaigns of the 1980s focused on political dimensions and social goals, this resolution also proposed to carry out trial construction of civilized cities, townships, and towns. Thus by the 1990s, when most research on cities in China focused on urban restructuring for economic development, CCP committees were articulating political thought work to develop policies for a new era of civilized places.

Tracing the origins of the civilizing imperative leads to an international encounter with emerging modernity in Japan, and its adaptations of European thought in French philosophy, in additions to similar ideological constructs in the former Soviet Union. While civilization or *wenming* 文明 appears scattered in early Chinese texts, its modern usage in China arrived in the late nineteenth-century exchange between China and Japan (Anagnost 1997, 80–1). As Geremie Barmé (2013, xvi–xvii) explains, Fukuzawa Yukichi (1835–1901) developed the idea of *wenming*, or *bunmen* in Japanese, in *An Outline of the Theory of Civilization*, published in 1875, based on François Guizot's 1828 *General History of Civilisation in Europe*. For Fukuzawa, civilization meant both material attainment and spiritual refinement. Thereafter, the idea of civilization as economic improvement and civility in society periodically appeared in Chinese social and political thought. After Mao's death and the end of the Cultural Revolution in 1976, the civilizational impulse began to feature again in Chinese political discourse. On National Day in 1979, PLA leader Ye Jianying called on the country not only to build "material civilization" (*wuzhi wenming* 物质文明) but also to reconstruct China's "spiritual civilization" (*jingshen wenming* 精神文明). This distinction retrieved the century-old debate over modernity in Asian society and repositioned it, for renewed guidance, in the reform era.

State interest in an ideology of civilization was also diffused to China in the process of the PRC's adoption of urban-industrial planning from the Soviet Union (Bernstein and Li 2010). A related concept, *kulturnost*, became articulated under Stalinism as a governmentalizing construct of social control. Vera Dunham (1990, 22) explains how kulturnost was "a fetish notion of how to be individually civilized." It was "admonitory and educative." Then it expanded from notions about basic manners, hygiene, and comportment to become "the self-image of dignified citizens" who "could now be models." "The notion of kulturnost had grown out of mores" and then "began to shape them, in accord with the regime's predilection for ponderous, monumental meshchanstvo." In the Soviet Union it had become a basis for the relationship between the regime and the lower middle class of soulless materialism. In contemporary China too its operational ambit interrelates the regime with the middle classes, whose support is effectively required in ongoing party-state leadership. Its recent iterations especially seek to entrain local officials in "civilized" administration.

The National Civilized City Program

The Spiritual Civilization Committee started setting standards and devising selection procedures for National Civilized Cities in the late 1990s. In the process, the Committee specified contexts of evaluation at levels of government in the territorial administrative hierarchy. Levels of government administration in China correspond to a specific area or mapped territory and exist in hierarchical relations under the central state. Official cities in China are administrative territories at three levels of rank: province-level city, prefecture-level city, and county-level city. Urban districts are subsidiary administrative areas in province-level and prefecture-level cities. The size of the four province-level cities – Beijing, Chongqing, Shanghai, and Tianjin – allows, under the civilized cities scheme, only their many subsidiary districts to be eligible for civilized city status. This complex territorial administrative organization reflects the history of establishing cities in the PRC under reform. Before the 1980s, territories at the prefecture and county levels were largely rural and the number of cities was fewer than 200 total. In the 1980s, China began reterritorializing prefectures and counties into prefecture-level and county-level cities, establishing over 450 new cities by the 1990s. In the context of territorial change for rapid urbanization, with hundreds of relatively new cities, the National Civilized City program exists to define, support, and guide urban development and social modernization.

In contrast to the dynamic textual basis of the National Civilized City Assessment System, the public records of National Civilized City honours are mere lists of place names. They provide only names and rank – that is, names of cities listed in order by administrative rank, from

10.4 "Establish a Civilized City, Construct a Harmonious Society." Xiamen, 2013.
Photograph by Carolyn Cartier.

prefecture-level and county-level cities to districts of provincial-level cities. In 2005 the Spiritual Civilization Committee named nine cities and three urban districts in the first group of National Civilized Cities – Dalian, Qingdao, Yantai, Baotou, Ningbo, Zhangjiagang, Xiamen, Shenzhen, and Zhongshan, and one district each in Beijing, Tianjin, and Shanghai. In 2009, eleven cities and three urban districts gained the honour, followed in 2011 by twenty-four cities and three districts, and thirty cities and four districts in 2015. From twelve cities and districts in 2005, the number of civilized cities and urban districts increased to over eighty by 2015 (People's Daily 2015). Among the original cities named in 2005, six cities and two districts retained the status through the ensuing evaluations, including Xiamen, a city opened to the world economy in 1980 as a special economic zone. Continuity and expansion of the National Civilized City program entrains its ideals, marks achievement of its standards, and makes commonplace how local officials contribute to producing uniformly "civilized" places. The program has also locally replaced red aesthetics with contemporary images, in pace with changing urban lifestyles (figure 10.4).

In 2005 the Committee also promulgated the National Civilized City Assessment System in the form of handbooks for each administrative level, and has since issued new editions in 2008, 2011, and 2015, or every three to four years. They instruct local governments how to apply for the scheme, beginning with screening and prequalification to enter the preparation and site evaluation process. Some prerequisites prevent a city's application. For instance, in the year preceding application or evaluation review, any rule violation or crime committed by a jurisdiction's Party secretary or mayor prevents application, as does any problem in the jurisdiction that theoretically could have been prevented by good government, such as accidental pollution of the water supply or a food security incident. Each jurisdiction has a Spiritual Civilization office that works with other branches of government to plan and implement activities that will meet goals and standards specified in the Assessment System. The 2011 edition added two new evaluation categories, one on adolescent development, category five, and one on social stability, category seven:

1. Incorruptible and efficient environment for government affairs
2. Democratic and fair legal environment
3. Fair and honest market environment
4. Healthy and progressive human environment
5. Socio-cultural environment for the healthy development of adolescents
6. Comfortable and convenient living environment
7. Safe and stable social environment
8. Ecologically sustainable environment
9. Solid and effective foundation for rural-urban policy implementation (My translation.)

The 2011 system manual features nine broad categories of evaluation whose targeted assessments, in multiple subsidiary categories, amount to over 100 items (National Civilized City Evaluation System Operation Manual, 2011). Category one, "incorruptible and efficient environment for government affairs," demonstrates the legacy of exemplary conduct in state-society relations while seeking to strengthen contemporary tenets of Party organization. It leads with continuing education for officials followed by administrative code of conduct, and measures for honest and industrious government officials. Each of these diversifies into two or more points of evaluation, under which fall detailed measurement goals. Criteria defining achievement in the category of administrative code of conduct ask officials to introduce expert advisory systems for public information, maintain a public service hotline with effective response mechanisms, and accept oversight of the news media and feedback from the general public. What is interesting about these, like the public feedback mechanism, is their global resonance under party-state control, exhibiting international standards of urban governance in an environment circumscribed by a variety of participatory limitations and surveillance capacities.

This condition is apparent in category two, "democratic and fair legal environment," in which democracy in China is "democratic centralism" whose conditions of participation take place substantially only within CCP ranks (Angle 2005; Tsang 2009). Its routinely unanimous outcomes are a main condition of the unitary party-state formation that is the PRC. Where most cities in China have a citizens' services centre – much like a city hall with information for local residents, including online resources – a democratic system with journalistic oversight remains only an aspiration. The Propaganda Department, whose official English-language translation was changed to Publicity Department in 2008, oversees and controls the media at large, including television and film. (The Ministry of Public Security controls the domestic Internet.) Freedom of the press in China, assessed by Reporters without Borders, has ranked at 176 out of 180 countries since 2015, which is four places lower than in 2010. A good example of increasing obstacles to press freedom is the 2013 "Southern Weekend incident," in which the Southern Weekend media group in Guangzhou encountered new restrictions from the state propaganda apparatus after a decade-long run of comparatively open national reporting (Ng 2013). The *Southern Weekend* newspaper had gained a reputation as a relatively innovative voice in the national media landscape. In the mid-2000s, Guangzhou, the provincial capital of Guangdong province bordering Hong Kong, was also a national leader in implementation of China's "open government" information initiative (Horsley 2007), which encourages governments to place new policies and regulations online. But Party documents are distinct from government documents, and many Party documents are internal only or classified and are not part of the government's open information initiative. Predictably, the records of the Spiritual Civilization Committee provide no open information about the assessment process or nullification of the award. When a city loses its title, its name simply disappears from the list.

In response to the National Civilized City program, city officials have become publicly involved in promoting their own interpretations of the civilized city as well as general and specific civilizing measures. For instance, in 2009 the mayor of Shenzhen, a prefecture-level city on the Hong Kong border, which was China's first special economic zone, launched a network of mayors to create civilized cities (Shenzhen News Network 2009). His leadership led to a series of television news talk shows featuring mayors and Party secretaries parsing interpretations of civilization for the contemporary city. These officials contributed to popularizing the idea through contemporary information streams, while also acting in the tradition of exemplars in society. By placing the conduct of Party officials at the top of the National Civilized City evaluation system, the Spiritual Civilization Committee exacts a governing relationship between the Party, local officials, and society. These officials as spokesmen, embodying and promoting the model of the National Civilized City, also demonstrate how it works as a tactic of government – a "dynamic form and historic stabilization of societal power relations" (Lemke 2002, 58).

Once a city receives the award, the process theoretically never ends – it faces future re-application and re-evaluation, or continual and periodic re-instantiation. Thus attainment of the honour is not a *fait accompli* but the opportunity to hold the title in a continual striving for improvement or "civilizing" over the ensuing three- to four-year period until the next evaluation. This capacity reaches its fullest optimization in the relationship between the National Civilized City award and the ranks of officialdom, in which the Spiritual Civilization Committee rescinds the honour in the event of any corruption incident by the Party secretary or mayor in an awarded city. This relation directly links the behaviour of local Party officials to the city's local and national standing, imparting particular awareness of regularity of action and rectitude in officialdom at large. It highlights the responsibilities of Party members, and Party oversight of government regulation and policy.

Even though reports of the evaluation process do not exist in the open record, it is possible to relate loss of the title to cases of corrupt officials. In several instances, the timing of corruption cases of ranking officials correlates with loss of civilized city status. In 2009 Qingdao and the Pudong district of Shanghai disappeared from the list. In 2011 Shenzhen failed to appear. In Shenzhen, in 2011, Xu Zongheng, the mayor who championed the civilized city idea, was removed from office for association with buying and selling official appointments (Wang 2011). The Shandong provincial capital of Qingdao lost its title in 2009 after authorities discovered that its Party secretary, Du Shicheng, took millions in bribes and awarded his mistress contracts for Olympics-related projects during preparations for the sailing competition of the 2008 Beijing Olympics (Caijing 2007). The Pudong district of Shanghai, site of the country's financial district of super-tall buildings, also lost its title in association with real estate corruption. The vice-mayor of Pudong from 2004–7, Kang Huijun, gained the moniker "director of real estate speculation" by giving land for development in exchange for bribes in cash payments and flats (Moore 2009). Yet like the periodic evaluation process instantiated in the model, a former civilized city can strive to improve to regain its award. In March 2013 Shenzhen became the first city to implement a civility law to make "uncivilized" public behaviour subject to fines (Shenzhen Daily 2013). In 2015 Shenzhen reappeared on the list.

Several categories of the Assessment System seek to promote development of the urban economy. Category four, "healthy and progressive human environment," features six main indexes, each of which includes up to eight points of evaluation that run the range from the Lei Feng model persona to creative industry: establishment of socialist values and morals, national education, development of cultural professions and cultural industry, activities and facilities for culture and sport, popularization of science, citizens' civilized behaviour, and volunteerism. The evaluation criteria require that public libraries and sports facilities provide free activity programs, and some museums and cultural and exhibition centres must provide free entry. Volunteerism is also a practice from the Mao era, when armies of volunteers supplied labour for rural development projects. Now, according to the manual, more than 90 per cent of the citizens of a city must demonstrate awareness of volunteer service organizations, and at least 8 per cent must participate in volunteer work. Urban volunteers are mostly women and students, and especially women retirees, since the official retirement age for women remains fifty-five years of age. University students with proficiency in multiple languages volunteer at major historical and cultural sites (figure 10.5).

The National Civilized City program also amalgamates existing policies from multiple government bureaus. With the second edition of Assessment System manual in 2009, the program incorporated standard measures of environmental quality. These reflect new directions in orthodox Party thought adopted in 2007. That year, Party chairman and president Hu Jintao proposed the concept of "ecological civilization" in his report to the Seventeenth National Party Congress of the CCP. With apparent awareness of hollow slogans, the "construction of ecological civilization is absolutely not rhetoric for chest thumping by officials in their speeches." It is "not a term the Party has coined just to fill in a theoretical vacancy in its socialism with Chinese characteristics" (China Daily 2007). The discourse shows continuity in Party thought, from socialism with Chinese characteristics to socialism with ecological characteristics, through "construction" (*jianshe* 建设), another keyword of the socialist lexicon. Jianshe appears widely in Mao-era slogans, such

10.5 "Volunteer Service Post." Xi'an, Mausoleum of the First Qin Emperor, 2013.
Photograph by Carolyn Cartier.

as "construct a dozen Daqing-style oil refineries, construct a dozen Dazhai-style communes." Tellingly, where *jian* in jianshe means to construct, the *she* in jianshe means, in addition to build, to establish and display – in effect, to show how the Party and the people together construct the nation. In jianshe is embedded the materiality of urban development and, now, tangible measures of ecological civilization, which must be attained to meet the evaluation criteria of the National Civilized City program.

Category eight of the National Civilized City evaluation system, "ecologically sustainable environment," is consistent with constructing improvement by applying technology to monitor and achieve environmental standards. It promotes, through three main indexes, greening the urban environment, environmental quality and management, and land-use-management standards. Each reflects the continuing role of urban and economic planning in urban modernization. Indexes for greening the urban environment calculate the extent of green space in residential areas and public park area per capita. Environmental quality management incorporates measurement tools to support mitigation of pollution, including urban noise pollution. The evaluation criteria for civilized air quality require meeting national air quality standards over 290 days per year. This goal sets the bar too high for some cities, though many urban governments have now expanded air quality monitoring of the standard indicators – PM2.5, O_3, CO, PM10, SO_2, and NO_2 – through an integrated, spatially distributed system of air monitoring and reporting stations. Where even in the mid-2000s information on particulate matter and air pollutants was not generally available in China, now anyone with access to the Internet can look up the air quality index of major cities in China.

The 2015 National Civilized City Assessment System manual introduced a new category and placed it first: "solid ideological and moral foundation." This new category of evaluation, focused on Party thought, displaced "incorruptible and efficient administrative environment." The first assessment item of the new leading category is "educate for ideals and belief." Its leading indicator is "study, publicize, and implement the spirit of the series of important speeches given by Xi Jinping." Its assessment criteria call for learning Xi Jinping's speeches as a "long-term political duty for the Party and the government." As a result, this priority appears in Spiritual Civilization Committee offices in all cities at all levels of government. In addition, the 2015 Assessment System consolidates multiple categories of the past versions into three:

1. Solid ideological and moral foundation
2. Favourable environment for social and economic development
3. Establishment of long-term general work mechanisms. (My translation.)

Category two of the 2015 manual encompasses most of the functional categories of prior Assessment Systems including cultural services and environmental quality standards (National Civilized City Assessment Manual, 2015). Category three, "establishment of long-term general work mechanisms," prioritizes "construction of spiritual civilization" through Party and government support, and assesses the city's master plan for elements of spiritual civilization construction. Thus among the three general categories of the 2015 version of the Assessment System, one prioritizes Party thought though explicit emphasis on the speeches of Xi Jinping, and another explicitly promotes the work of the Spiritual Civilization Committee through general city planning. Under the leadership of Xi Jinping, who in 2012 became general secretary of the CCP, president, and chairman of the Central Military Commission, the Party has sought to consolidate and strengthen power relations. In 2018, the Party Central Committee removed the two-term or ten-year limit on the offices of president and vice-president, allowing Xi to continue. The National Civilized City Assessment System reveals how ideological planning widely anticipated and promoted the Xi Jinping era.

Conclusion: A Hybrid Model of Standard Urban Governance

The PRC has been developing hundreds of new cities under reform, requiring a comprehensive approach to urban transformation. Still, many of its policy forms are neither readily visible to observers nor directly comparable to approaches in international urban planning. The National Civilized City program articulates measures for achieving multiple urban development goals by promoting coordinated modernization among spheres of government, society, the economy, the environment, and culture (as conceived by the state). It promotes the replicability of governing standards and seeks to entrain both officials and citizens in reproducing a CCP-defined urban modernity. Its expansive brief underscores how the program has grown beyond a set of political-linguistic techniques for socializing citizen conduct in ideas about population quality, to become a general governing approach for improving the urban environment. The role of Party officials in upholding the standards of the civilized city through exemplary conduct underscores the program's capacity for meta-governance.

As a Party program, with continuing use of red slogans, the National Civilized City is also a commercialized form of socialist mobilization, or agitprop, linking Maoist traditions of mobilizing the masses and model strategies with contemporary urban planning. The civilized city's formal and aesthetic roots lie in Mao-era political campaigns, employing model people and places to socialize results. Today, these results tally up in response to the promotion of standard governing measures. Instead of promoting a model based on a particular place, as in the Mao era, the National Civilized City program operationalizes the style of a political campaign to promote a standardized model of urban management for all cities. Where the program continues the tradition of mobilizing volunteers, it combines techniques of social control with initiatives for generating citizen participation in maintaining quality of life in the city. Propaganda continues, while building the new modern city incorporates governing measures that strike new bargains with urban citizens. The model of the National Civilized City establishes a set of rules, regulations, and standards that transmit current ideals of urban modernization, officially guided. The program's recent revisions seek to instantiate in the city new relations of power between the party-state and society. This new socialist city represents a version of world-class city with Chinese characteristics.

REFERENCES

Anagnost, Ann. 1997. *National Past-Times: Narrative, Representation, and Power in Modern China*. Durham, NC: Duke University Press.

– 2004. "The Corporeal Politics of Quality (*suzhi*)." *Public Culture* 16, no. 2: 189–208.

Angle, Stephen C. 2005. "Decent Democratic Centralism." *Political Theory* 33, no. 4: 518–46.

Bakken, Børge. 2000. *The Exemplary Society: Human Improvement, Social Control, and the Dangers of Modernity in China*. Oxford, UK: Oxford University Press.

Barmé, Geremie. 2013. "Engineering Chinese Civilisation." In *China Story Yearbook 2013: Civilising China*, edited by Geremie R. Barmé and Jeremy Goldkorn, i–xxix. Canberra: Australian National University.

Bernstein, Thomas, and Hua-Yu Li, eds. 2010. *China Learns from the Soviet Union*. Lanham, MD: Roman & Littlefield.

Brady, Anne-Marie. 2008. *Marketing Dictatorship: Propaganda and Thought Work in Contemporary China*. Lanham, MD: Rowman and Littlefield.

Caijing. 2007. "Top Qingdao Official Removed, Investigated Amid Soaring Housing Prices." *Caijing.com.cn*, 18 January. http://english.caijing.com.cn/2007-01-18/110031980.html

Cartier, Carolyn. 2013. "Building Civilised Cities." In *China Story Yearbook 2013: Civilising China*, edited by Geremie R. Barmé and Jeremy Goldkorn, 265–86. Canberra: Australian National University.

– 2016. "Governmentality and the Urban Economy: Consumption, Excess and the 'Civilized City' in China." In *New Mentalities of Governance in China*, edited by Elaine Jeffreys and David Bray, 56–75. Abingdon, UK: Routledge.

CCP Central Committee. 1996. *Resolution Concerning Several Important Questions for Strengthening Socialist Spiritual Construction* [booklet]. Beijing, China: People's Publishing House.

China Daily. 2007. "Ecological Civilization." *China Daily*, 24 October. http://www.chinadaily.com.cn/opinion/2007-10/24/content_6201964.htm

Deng, Xiaoping. 1982. "Opening Speech at the Twelfth National Congress of the Communist Party of China." Speech given 1 September, Beijing. http://en.people.cn/dengxp/vol3/text/c1010.html

Dunham, Vera. 1990. *In Stalin's Time: Middleclass Values in Soviet Fiction*. Durham, NC: Duke University Press.

Friedmann, John. 2005. *China's Urban Transition*. Minneapolis: University of Minnesota Press.

Funari, Rachel, and Bernard Mees. 2013. "Socialist Emulation in China: Worker Heroes Yesterday and Today." *Labour History* 54, no. 3: 240–55. https://doi.org/10.1080/0023656x.2013.804269

Hoffman, Lisa. 2011. "Urban Modeling and Contemporary Technologies of City Building in China: The Production of Regimes of Green Urbanisms." In *Worlding Cities: Asian Experiments and the Art of Being Global*, edited by Ananya Roy and Aihwa Ong, 55–76. Chichester, UK: John Wiley and Sons.

Horsley, Jamie P. 2007. "China Adopts First Nationwide Open Government Information Regulations." *Freedominfo.org*, 9 May. http://www.freedominfo .org/2007/05/china-adopts-first-nationwide-open-government-information -regulations/

Huang, Yasheng. 2008. *Capitalism with Chinese Characteristics: Entrepreneurship and the State*. Cambridge, UK: Cambridge University Press.

Jeffreys, Elaine M. 2012. "Modern China's Idols: Heroes, Role Models, Stars and Celebrities." *PORTAL Journal of Multidisciplinary International Studies* 9, no. 1: 1–32. https://doi.org/10.5130/portal.v9i1.2187

Larson, Wendy. 2009. *From Ah Q to Lei Feng: Freud and Revolutionary Spirit in Twentieth-Century China*. Stanford, CA: Stanford University Press.

Lemke, Thomas. 2002. "Foucault, Governmentality, and Critique." *Rethinking Marxism* 14, no. 3: 49–64.

Leung, Angela K.C. 1993. "To Chasten Society: The Development of Widow Homes in the Qing, 1773–1911." *Late Imperial China* 14, no. 2: 1–32. https:// doi.org/10.1353/late.1993.0006

Lim, Kean Fan. 2014. "'Socialism with Chinese Characteristics': Uneven Development, Variegated Neoliberalization and the Dialectical Differentiation of State Spatiality." *Progress in Human Geography* 38, no. 2: 221–47.

McGee, Terry, George C.S. Lin, Mark Wang, Andrew Marton, and Jiaping Wu. 2007. *China's Urban Space: Development under Market Socialism*. London, UK: Routledge.

Moore, Malcolm. 2009. "Chinese Skyscraper Official Jailed for Taking £2m in Bribes." *The Telegraph*, 4 February. https://www.telegraph.co.uk/news /worldnews/asia/china/4510763/Chinese-skyscraper-official-jailed-for -taking-2m-in-bribes.html

National Civilized City Assessment System Operation Manual (全国文明城市 测评体系操作手册). 2011. Beijing, China: Central Civilization Office (中央文 明办).

– 2015. Beijing: Central Civilization Office (中央文明办).

Ng, Teddy. 2013, "Outrage at Guangdong Newspaper Forced to Run Party Commentary." *South China Morning Post*, 4 January. http://www.scmp .com/news/china/article/1119378/outrage-guangdong-newspaper-forced -run-party-commentary

Penny, Benjamin. 2013. "An Exemplary Society." In *China Story Yearbook 2013: Civilising China*, edited by Geremie R. Barmé and Jeremy Goldkorn, 152–77. Canberra: Australian National University.

People's Daily. 2015. "National Civilized City Now Includes 16 Provincial Capitals." [In Chinese]. *People.cn*, 5 August. http://politics.people.com.cn /ywkx/n/2015/0805/c363762-27416022.html

Shapiro, Judith. 2001. *Mao's War against Nature*. Cambridge, UK: Cambridge University Press.

Shenzhen Daily. 2013. "Civility Law to Take Effect March 1." *Chinalawinfo*, 24 January. www.lawinfochina.com/Search/DisplayInfo. aspx?lib=news&id=10693

Shenzhen News Network. 2009. "Mayor Xu Zongheng Guest on Shenzhen News Network to Talk about the Civilized City." [In Chinese]. *Sohu.com*, 4 June. https://media.sohu.com/20090604/n264326500.shtml

Solé-Farràs, Jesús. 2008. "Harmony in Contemporary New Confucianism and in Socialism with Chinese Characteristics." *China Media Research* 4, no. 4: 14–24.

Tang, Wing Shing, and Him Chung. 2002. "Rural–Urban Transition in China: Illegal Land Use and Construction." *Asia Pacific Viewpoint* 43, no. 1: 43–62. https://doi.org/10.1111/1467-8373.00157

Tsang, Steve. 2009. "Consultative Leninism: China's New Political Framework." *Journal of Contemporary China* 18, no. 62: 865–80. https://doi .org/10.1080/10670560903174705

Tsing, Anna Lowenhaupt. 2000. "Inside the Economy of Appearances." *Public Culture* 12, no. 1: 115–44. https://doi.org/10.1215/08992363-12-1-115

Wang Xianwei. 2011. "Just the Tip of an Iceberg for Official Graft." *South China Morning Post*, 2 May. http://www.scmp.com/article/966593 /just-tip-iceberg-official-graft

Wilhelm, Katherine. 2004. "Rethinking Property Rights in Urban China." *UCLA Journal of International Law and Foreign Affairs* 9, no. 2: 227–300.

Zhao, Jingzhu. 2011. *Towards Sustainable Cities in China: Analysis and Assessment of Some Chinese Cities in 2008*. New York, NY: Springer.

11 Property Relations and the Politics of the Suburban Living Place in the Post-Communist City: Transition Stories from Tirana, Albania

MARCELA MELE AND ANDREW E.G. JONAS

One well-known feature of capitalist forms of urban development in North America and western Europe is the politics of the suburban living place, the spatial appearance of which seems to be separate from class and political tensions arising in urban society at large (Cox 2002). Recognized features of such a politics include homeowners' efforts to secure use values in the communities where they live (e.g., access to good local schools) and thereby enhance exchange values (Logan and Molotch 1987), exclusionary political practices originating in the residential neighbourhood or suburb (Cox and Johnston 1982), and demands for local improvements in education and other public services (e.g., water and sewage, street maintenance, etc.) – demands that might eventually lead to the creation of separate local jurisdictions in the suburbs (Cox and Jonas 1993). It might be expected that suburban areas in post-socialist states undergoing a transition to capitalist property relations likewise experience similar tensions around the living place, which in turn give rise to claims for territorially separate territorial structures of suburban government or governance.

There is growing evidence from former socialist and communist countries that a global convergence is indeed occurring around forms of suburban governance that bear some resemblance to the Anglo-American model of polycentric urbanism (Hamel and Keil 2015; Hirt 2012). However, in certain contexts post-socialist patterns and processes of suburban governance *diverge* from the Anglo-American model (Hirt and Kovachev 2015). For example, the absence until quite recently of formal local government with such powers as taxation, property registration, land use planning, and so forth has meant that suburban areas in some former communist countries have been colonized by informal systems of planning, regulation, and development. So perhaps divergence should neither surprise nor overly concern us. In fact, scholars of comparative urbanism have become increasingly sceptical of attempts to transpose models and theoretical frameworks derived from cities in the Global North like Chicago or Los Angeles to "other" urban contexts, especially those in the Global South (Robinson 2006; Sjöberg 2014). Instead, such scholars are motivated to examine and reveal the lived experiences, practices, institutions, and livelihoods of those who inhabit the world's diverse urban (and suburban) spaces, irrespective of prior theoretical claims or empirical generalizations based on cities in North America and western Europe (Jonas, McCann, and Thomas 2015). The premise of this chapter is that there are many more "transition stories" (Pickles and Smith 2005) still to be revealed about unfolding patterns and processes of suburban development, which might help to shed further light on an emergent politics of the suburban living place that is specific to different post-socialist settings.

A case in point concerns the transition from communism to capitalism in Albania, from 1991 to the present day. Albania represents an intriguing case to examine in a book about socialist and post-socialist urbanisms. From 1945 until 1990, the country adopted an extreme system of communist rule that took on a specific form of "national Stalinism" (Judt 2010, 430). Under the regime of President Enver Hoxha (1944–85), the Albanian state nationalized land, abolished private property relations, and prohibited urban development, resulting in a greatly impoverished and predominantly rural society. Surrounded by imagined enemies, Albania remained an ethnically closed and geopolitically isolated country until the collapse of communism at the start of the 1990s (Zickel and Iwaskiw 1994). Following the liberalization of trade and the opening of its borders in 1991, Albania, along with its former socialist European neighbours, began to exhibit signs of convergence around neoliberal processes and patterns of urban development. However, the country's unique encounter with communism has left a deep imprint on urban society and suburban governance. This is especially the case in Tirana, the capital city, where suburbanization has been profoundly shaped by conditions of informality, rapid rural-to-urban migration, competing claims on property, struggles for political citizenship, and at times outright conflict.[1]

Highlighting the case of Kamza (shown as *Kamëz* in figure 11.1), a municipality located immediately to the northwest of the capital city, Tirana (or *Tiranë*), this chapter examines how one urban settlement in Albania rapidly transitioned from a rural commune in 1991 into a separate suburban municipality in 1996, which to this day confronts the many and varied challenges of rapid urbanization under conditions of informality (Pojani 2013). It draws on field research undertaken by one

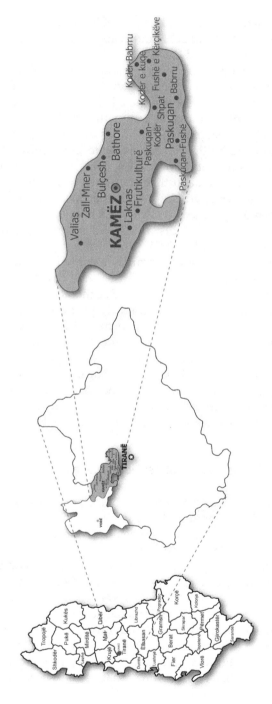

11.1 Location of Kamza in relation to Tirana and districts of Albania. Figure credit: Marcela Mele, 2019.

of the authors (a native of Albania) as part of her doctoral dissertation undertaken between 2008 and 2011, and is augmented by data and observations collected in the field in 2014.

The research involved a mix of observational fieldwork, ethnography, archival analysis, and face-to-face-interviews. A total of forty-nine semi-structured interviews were conducted with individuals grouped into the following categories: recent migrants to Kamza (five interviews), long-standing residents (ten), former owners whose property had been expropriated by the communist state (five), local public officials and politicians (twenty-six), and developers (three). Interviews were conducted in Albanian and then translated into English (sample quotes from the transcripts are reported here). Interviews were structured around themes designed to elicit local knowledge about property ownership, access to services, livelihood strategies, and voting behaviour (for details, see Mele 2011). They were supplemented with informal (i.e., noted down but not formally recorded and transcribed) conversations with residents in streets, buses and bars in order to garner a richer portrait of suburban lives and livelihoods.

We found that in the Albanian context, post-communist suburban governance reflects a complex conjuncture of flows (e.g., internal rural-urban migrations), fluxes (rapid and informal urbanization, especially around Tirana), and fixities (e.g., claims for property rights in and around the suburban living place). The establishment of territorially separate institutions of suburban governance has been supported by international organizations including the World Bank and NGOs, and is sustained through a variety of livelihood strategies constructed in the suburbs.

Urbanization in a Post-Communist State: Albania in Context

For the past two decades or more scholars of the post-socialist city have engaged closely with the profound changes occurring throughout Eastern Europe, a region that continues to deal with the political legacy of urban development patterns and processes established under different models of socialism and communism (Andrusz, Harloe, and Szelenyi 1996). Following the end of the Soviet system of central planning in 1990, the region's transitional economies were each in turn exposed to broadly similar processes of economic globalization and neoliberal state restructuring (Pickles and Smith 1998). But urban social, economic, and political outcomes varied markedly throughout the region, reflecting differences in state institutions, economies, and local civil society (Stenning 2003; Smith 2007). Given such national variations, Sýkora (2008) makes the case for examining place-specific trajectories of urban development in the post-socialist states of Europe. In this light, the former communist state of Albania represents an intriguing context for enriching our knowledge of post-communist and post-socialist urbanism and suburbanism.

At the turn of the twentieth century, Albania was a rural country with a scattering of urban settlements based around local bazaars, mosques, and other public institutions, many a legacy of an earlier period of Ottoman rule. As the country became increasingly caught up in European imperial rivalries, Ahmet Muhtar Zogolli or Zog I (president, 1925–8, and then king, 1928–39) sought to transform the largest urban centre, Tirana, into a European capital. Impressive boulevards and grand administrative buildings were added to the existing urban fabric, followed by further transformations under Italian occupation by Benito Mussolini's Fascist regime in the 1930s. At the start of the Second World War, Tirana had a population of 35,000 in a country containing about 1 million people (Mele 2011, 43). A master plan was developed for Tirana that sought to reflect Mussolini's aspirations for empire and urban spectacle.

After the defeat of Fascism and the end of the war, Albania embraced an extreme form of nationalist Stalinism, which brought an end to the monarchy, abolished private property, and nationalized industry, agriculture, and urban real estate (Abitz 2006). In 1946, agricultural land and assets were collectivized by the state and placed under the control of cooperatives and state farms with a view to achieving self-sufficiency. By 1976, all land in Albania had been nationalized and the state even exercised control over moveable goods such as furniture, motor vehicles, and luxury household items. Moreover, successive five-year plans prioritized industrialization over urbanization. Urban growth was restricted by strict controls on rural-to-urban migration and the designation of urban boundaries in the form of "yellow lines" identified in regional plans (Sjöberg 1992). These policies were highly effective in restricting urban development. By 1990, only nine out of Albania's thirty-six designated urban centres had populations greater than 25,000 in a country of about 3 million people (Mele 2011, 46).

Despite such restrictions on urban growth, Tirana's population grew from 60,000 inhabitants in 1945 to around 226,000 in 1990 (Mele 2011). As the country's main political, industrial, educational, and cultural centre, Tirana not only benefitted from the communist state's industrialization policies but also from "diverted migration" (Sjöberg 1992). This refers to when migratory flows heading for a particular destination (such as the capital city) are diverted to nearby destinations, in this case Tirana's rural hinterland. It appears that migrants, who were unable to obtain permission from the state to move their residence to Tirana proper, managed to migrate to one of the rural cooperatives or state farms in the immediate vicinity of the city. As these areas began to attract higher population densities, the state responded by altering their administrative designation from rural to urban.

The collapse of communism in 1990–1 witnessed a rapid acceleration of rural-to-urban migration within Albania and the explosive development of Tirana's rural periphery (Aliaj, Lulo, and Myftiu 2003; Abitz 2006). During the communist era (1945–91) Tirana's urban limits were defined by a "yellow line" marking the official boundary between the city and outlying collective farms and agricultural zones. With the lifting of the urban growth restrictions in 1991, areas beyond the "yellow line" began to attract rapid new development. During a period known as *kioskism* (1991–7), migrants seized control of land at the edge of Tirana and engaged in petty forms of capitalist production and exchange, including growing produce for local markets, using their homes as restaurants and retail outlets, and street vending. In Tirana itself grand public spaces laid out in the 1920s, including boulevards, parks, and squares, became cluttered with shops and a diverse assortment of other quasi-legal enterprises (Sluis and Wassenberg 2003). Kioskism came to a dramatic end following the collapse of Albania's pyramid investment schemes in 1997. However, Tirana's periphery continued to attract migrants from rural districts, especially in the north and east of the country. By 2011, Tirana's population had grown to more than 600,000 compared to the national population of 3.2 million, which because of international migration had declined from its peak of 3.3 million in 1997 (Mele 2011, 55).

In the District of Tirana, spacious private residences and farms – some a legacy of pre-communist patterns of construction and land ownership – became interspersed with densely packed informal housing. Initially these informal settlements had no formal jurisdictional status, nor did the migrants settling in such communities necessarily possess a sense of a shared political identity based upon ethnicity/region of origin. Such settlements became physically attached to Tirana, effectively creating a continuously built-up urban area. In the absence of a corresponding urban jurisdictional status, the informal settlements often lacked basic infrastructure and services such as paved streets, schools, community facilities, and a reliable supply of electricity. The informal and fragmented nature of urban development around Tirana both during and immediately after the kioskism period soon led to efforts to establish a more formal system of suburban governance exemplified by the case of Kamza.

Building Suburban Governance: The Case of Kamza Municipality

During the communist era, the area known as Kamza (or *Kamëz* in figure 11.1) was a state farm producing agricultural goods, food, and livestock for the capital city, Tirana. Despite its attractiveness to migrants, Kamza avoided an urban designation. After the collapse of communism in 1990, Kamza and the surrounding area were settled illegally by migrants from other regions of Albania. Tsenkova (2010, 79) suggests that up to 90 per cent of the dwellings in and around Kamza are illegal constructions. Kamza retained its rural administrative status until 1996, at which point it was officially incorporated as a municipality to become part of the Greater Tirana District (hereafter GTD), which assumed overall responsibility for regional planning.[2] The boundaries of the Municipality of Kamza today encompass two established urban centres, Kamza Centre and Frutikulture, and three rural communities, Bathore, Bulcesh, and Zallmner, which had formerly been part of the state farm (figure 11.1). Responsibility for planning and administration within Kamza is divided between Kamza Centre, Bathore, and a further five administrative divisions. Kamza and its constituent communities grew from less than 10,000 residents in 1990 to more than 90,000 in 2010 (Mele 2011, 123). In 2015, the enlarged Municipality of Kamza was estimated to have approximately 104,000 residents.

The growth of informal settlements has been accompanied by demands for suburban political autonomy and growing political tensions within the GTD, which have often assumed a territorial form. One source of ongoing tension relates to whether or not Kamza (or some of its administrative functions) should be annexed to, or consolidated with, the City of Tirana. Municipal annexation and metropolitan consolidation are established practices in North America, where the motives include cheaper services for residential developers, enhanced tax revenues for municipal governments, and economic development subsidies for industrial branch plants attracted to the urban periphery (Jonas 1991). In the Albanian context, however, *annexation* is a politically controversial term because of its historical association with occupation (e.g., by Fascists both before and during the Second World War) and the dispossession of the rural peasantry and landowners (e.g., under Ottoman rule and later under communism). Therefore most public officials shun the use of the term *annexation*, which leads to a lack of open public debate and discussion of the benefits (or otherwise) of amalgamating communes and municipalities, even though such strategies might be viewed as fiscally rational and desirable from an urban development perspective. Nevertheless, over the years Kamza has gradually consolidated its constituent local administrative units. The City of Tirana itself has expanded to incorporate at least eleven administrative units (known as *mini-bashki* or small municipalities), in the process enlarging its population and accruing revenue for undertaking significant investments in urban infrastructure (e.g., the construction of a major ring road around Tirana and the widening of the city's streets and boulevards).

A further source of tension involves the need to strike a balance between building up suburban governance capacities, on the one hand, and securing state support for urban development, on the other. Since the collapse of communism, independent self-government on the local level has been the guiding principle for fostering democracy throughout Albania's urban settlements (Dedja and Brahimi 2006). However, there are ongoing tensions between the national state and local government about the level and scope of decentralization in taxation and capital investment. In part, these tensions reflect processes of uneven regional development. Throughout the 1990s, the country's remoter rural regions rapidly depopulated, thereby placing enormous pressures and demands for urban services on the GTD, which was the preferred destination for a majority of internal migrants.

In other European states, uneven regional development has been both a cause and the consequence of the hollowed-out local state and the emergence of new structures of urban and regional governance (Goodwin and Painter 1996). The early phases of territorial decentralization in Albania involved the enhancement of local government capacities in the cities alongside efforts to strengthen local government, a process that was mediated by international non-governmental organizations (NGOs) and the World Bank. However, the pace of land development in peripheral parts of Tirana soon overstretched the fiscal and administrative capacities of nascent units of urban government, and it was not until 1996 that a separate municipal authority was formally established for Kamza, which assumed responsibility for public administration and the collection of taxes for local services. Accordingly, urban government in Albania has often been found wanting in capacities sufficient to address the profound challenges of haphazard urbanization, informality, and lack of basic services, notably roads, water, and electricity, especially in recently settled suburban areas (Pojani 2013).

One of our respondents, a migrant living in the Bathore area of Kamza, described how in the absence of urban planning regulations residents initially took matters into their own hands:

> The newcomers settled in the Kamza area, especially in Bathore, with no infrastructure and urban planning. The Bathore area was a hilly agricultural zone with open free lands. Initially the houses were widely dispersed but developed with such a high speed that conflicts quickly emerged. I have some specific examples: two or three householders had [built] their houses close to each other. Some others have their houses a bit further apart. But all of them [had access to] a common narrow lane. One day, one of the households blocked the entry way [to the lane], leaving the other houses without access to the lane. (Interview with migrant, Bathore)

During the kioskism period, NGOs intervened in areas like Bathore to establish structures of public participation, which were informed by models of "good governance" circulated by international organizations such as the United Nations and the World Bank (Co-PLAN 2002a). One such NGO, Co-PLAN, became involved in the informal settlements of Kamza, strengthening local society and governance through programs of public participation and engagement (Hartkoorn 2000). Co-PLAN initially focused on developing participatory governance alongside civil society. Later it assisted in the transfer of capacities for land use planning, service delivery, and waste collection from the national and regional levels to the newly formed Kamza Municipality (Co-PLAN 2002b). Thanks in part to the efforts of Co-PLAN, Kamza was soon regarded as a favoured partner in governmental reform and modernization in Albania, attracting central state and international grants for major infrastructure projects, which transformed the built environment of Kamza Centre as well its informal settlements (Co-PLAN 2002a; Council of Europe 2008; Mele 2011).

The early 2000s witnessed a boom in construction activity, resulting in the proliferation of multi-storey apartments and commercial buildings, both in Kamza and the rest of the GTD. In general, however, large-scale infrastructure projects, such as the ring road, were earmarked for the City of Tirana because of its status as the national capital and the close political ties existing between the city government and the ruling socialist administration at the national level. This created some tensions with the local municipalities. Responding to ongoing local demands for improvements in services in areas such as Kamza, the Albanian state undertook further legal and institutional measures to transfer responsibility for the value-added tax, local taxation, water supply, and sanitation from the central government to the municipalities (Dedja and Brahimi 2006). These measures, in turn, spurred efforts to register and legalize properties in the informal settlements to create a more reliable and efficient tax collection system. Kamza Municipality's growing capacity to issue construction permits and collect property taxes significantly enlarged its capital budget over subsequent years.[3]

Throughout the GTD the legalization of informal property along with the expansion of organized large-scale construction activity have created a context for competing and conflicting local accumulation strategies around the suburban living place (see Cox and Jonas 1993). These strategies range from the further expansion of commercial, residential, and entertainment activities in the informal settlements to more organized and systematic building activities in the legalized areas, producing densely packed areas having a mixture of small-scale family homes, retail outlets, and offices, along with well-planned spaces with high-rise (five or six storey) residential apartments. In Kamza, for example, new residential and commercial developments vary not only in size and architectural character but also in the levels of investment in supporting street infrastructure (roads, sidewalks, sewerage, electricity, green space, etc.), a factor that creates the impression of a colourful patchwork quilt of built environment forms around the suburban living space (figures 11.2 and 11.3).

Property Relations and the Politics of the Suburban Living Place

Writing in 2007, the authors of a World Bank report (2007a, n.p.) argued that "secure property rights along with an efficient and transparent land management regime are fundamental for creating well-functioning land and property markets in Albania." Yet for many ordinary Albanians "secure property rights" is a profoundly contested idea, not least in the peripheral urban settlements of Tirana. Following the end of communism, in 1991 the Albanian state introduced legislation to establish the legal basis for private property relations. However, the process of legalization soon created new social divisions and tensions. For example, Land Law 7501 restored ownership rights to families that resided on the cooperatives and state farms at the time of the law's passage but did not provide for properties to be restituted to families originally affected by collectivization in 1945. Moreover, the law was intended to redistribute all collectivized land to former cooperative members on an equal per capita basis but other rural residents, many of whom were not members, were also awarded land, if in smaller quantities. While such laws facilitated the privatization of agricultural land in close proximity to cities, remoter forests and pastures by and large remained state owned (de Waal 2004). Without effective state regulation of property, corruption, illegality, and self-regulation became widespread practices, dictating property relations and livelihood strategies in both rural and urban areas immediately following the collapse of communism.

11.2 Mixed commercial and residential developments and poorly maintained street infrastructure in the centre of Kamza. Photo credit: Marcela Mele, 2019.

11.3 New apartments and well-maintained street infrastructure on the edge of Kamza. Photo credit: Marcela Mele, 2019.

In the meantime, the legalization of property relations and the regulation of land-use planning at the urban fringe continued to occur incrementally, frequently prompted by exhortations and interventions by neoliberal institutions such as the World Bank (Kelm 2002; World Bank 2007b; Childress 2006). A formal process of compensation and property restitution was not initiated until 2004 when the Albanian state government initiated a program to address the problem. Law 9304, "On Legalization and Urban Planning of Informal Zones," formally identified such zones as a prelude for transitioning them into properly functioning urban districts (Andoni 2007a, b). Mechanisms were put in place to legalize informal residential and commercial properties in these zones, and leasing rights were given to their occupants (World Bank, cited in Tsenkova 2010).

Nevertheless, reforms to property relations often failed to prevent individuals from further building on land without legal authorization. Because of delays in property restitution actions, the illegal construction of buildings on the fringe of cities became widespread, resulting in conflicts between new migrants and long-standing residents having competing claims in the suburban living space (Felstehausen 1999). Kamza became a flashpoint for struggles around the restitution of property rights within the GTD. Between 1995 and 1998 there were a series of clashes between the state police and local residents in the informal settlements (Aliaj et al. 2003). In response to growing conflict in the area, and under pressure from international institutions and NGOs, the state eventually "forgave and justified" illegal building activity in Kamza. Property legalization soon became a mechanism for capturing local votes in the suburbs, often exacerbating political schisms between Kamza and the City of Tirana (Mele 2011).

One reason for the emergence of such tensions around the suburban living place in Kamza (as well as other informal settlements) is the uneven social division of property relations, which gives rise to competing claims in respect to ownership of, or access to, land and property. Social divisions in the suburbs are not specific to Albania; yet in this context they essentially revolve around divisions between legal owners (long-standing residents), outside landholders (including the state), developers who acquired property prior to its legalization, and migrants who have occupied property in the informal settlements. These differences appear in table 11.1, which describes salient categories of property relations according to field observations and data gathered between 2008 and 2010.

Social divisions around post-communist property relations in Albania have underpinned not only a variety of local accumulation strategies but also local livelihood practices around the living place.

11.1 Social categories of property relations in suburban areas of Tirana

Property relations	Salient characteristics
Legal owners	The state
	Private owners as defined in Albania's Land Laws 7501, 8053, and 7698
Illegal or informal buyers	This includes buyers who bought land informally from legal private owners, as well as those who subsequently bought land from the buyers. They possess no legal documentation certifying these transactions or their ownership.
Illegal owners	This includes people who have illegally settled on: (1) state-owned land or; (2) land owned by private individuals. Neither of these groups possess documentation certifying ownership.

Source: Compiled by authors.

Such practices range from rights claims for a place of abode (i.e., land or property) to various acts of self-regulation on the part of long-standing residents and property claimants. The following interviews drawn from field research give some sense of their precarious nature, and the sometimes unresolved claims for a suburban living space on the part of migrants in Kamza and its constituent communities:

> When I moved to Bathore I asked one of my former neighbours [who] settled here a few years ago. He had some extra land and gave me 500 square metres ... In the very first year I built on my own a simple portable house. I went back to my village, brought my family back and all together built the new concrete house, while living in the hut. I provided enough money from my part-time work in construction and some loans from my relatives. (Male migrant, Bathore)

> We moved to Kamza in 2006 and rented a house quite cheap. My husband and I rent a shop along the sideway. We sell second-hand clothes. We struggle and can't even afford many new items [of clothing]. (Female migrant, Kamza Centre)

To secure livelihoods and claims to a living space, migrants initially resorted to various forms of self-exploitation, such as drawing upon the resources of the extended family (e.g., remittances) to undertake street improvements and provide basic services. Only later did they become conscious of the need to claim entitlement as a precondition for securing essential services:

> When my family moved here there were only two or three families. While the house was being built I stayed in my brother's house. I have a brother

who has a house close to me. My family and I stayed one year with him. We have a road, but it is not asphalted. Three [local] elections have since occurred and [the politicians] keep promising us a road, sewage system, and energy; but nothing has happened up until now. As a neighbourhood we did some of the [road] work on our own efforts ... Even for the drinking water system, we collected money from four or five families. (Interview with female migrant, Kamza)

After 2000 our attitude about property started to change. Before we used to build a house on empty land and we dare not ask "what's next?" Recently, we started to understand that if we don't get entitlement we can't have access to land transaction, infrastructure, electricity, mail, etc. (Interview with male migrant, Kamza)

Although many migrants in Kamza have resided in the informal settlements for the best part of two decades, the failure to secure property rights remains for some an obstacle to securing local voting privileges and hence also a sense of suburban citizenship.[4] The state agency responsible for securing such rights is the Agency for Legalization and Urbanization of Informal Settlements or ALUIZNI (Tsenkova 2010). ALUIZNI has struggled to deal with the exceptionally high volume of claims and the complexity of issues surrounding entitlements to property (Andoni, cited in Tsenkova 2010). In 2007–8, a total of 12,571 claim files were submitted to ALUIZNI to secure entitlement to property. Although processing such claims is an important part of the process of property registration, there have been tensions between the local authorities in Kamza and ALUIZNI over whether legalization should take precedence over developing a municipal land use plan. Significant areas of Kamza have outstanding entitlement claims (table 11.2) that cast a shadow over ongoing efforts to formalize local planning, public administration, and citizenship around the suburban living place.

Conclusion

The form of post-communist suburban governance in Albania can be understood as a particular response to ongoing tensions associated with various livelihood practices in the suburban space. Such tensions include: migrants struggling to obtain material resources and claiming access to a place to live; residents and former property owners making claims for recognition of their entitlement rights; individual land and property speculators taking advantage of the uncertain legal status of suburban land; and residents and communities self-financing,

11.2 Informal zones and outstanding property rights claims in Kamza

Informal zones	Surface area in ha	Number of buildings	Number of buildings with outstanding claims
Zones 1, 2, 3, and 4 in Bathore and Bulçesh	240.7	2,958	2,803
Zones 5, 6, and 7 in Bathore	268	2,450	2,311
Kamez	536	2,900	2,800
Laknas	402	1,330	1,226
Valias	329	1,545	1,317
Frutikulture and Zall-Mner	208.7	1,825	1,768

Note: Compiled from local newspapers and property records; zones and areas from Co-PLAN 2002b.

self-providing, and self-maintaining essential services. In light of these tensions and their associated outcomes, it is dangerous to reduce suburban governance in post-communist Albania to a simplistic model of transition from state control to private property relations and suburban political autonomy. Instead, the territorially separate institutions of suburban governance have been produced and sustained through a variety of informal and formal livelihood strategies along with claims for a right to an abode, especially for migrants.

Whether the Albanian experience of post-communist suburban governance is sustainable and generalizable to other contexts is open to further discussion and investigation. Until quite recently, official digital maps of the GTD did not show the jurisdictional boundaries of Kamza in their entirety, suggesting that in the eyes of state authorities the system of suburban governance had not yet matured sufficiently for peripheral settlements like Kamza to be recognized as territorially separate local jurisdictions. Moreover, following the 2007–8 global financial crisis and changes in national government, tensions between the Albanian state and local government have re-emerged. Any gains in suburban political autonomy immediately after transition in the early 1990s might soon be compromised by an unstable global regulatory context and ongoing pressures on the Albanian state to recentralize certain functions relating to urban planning, public services, and taxation. Across Tirana and the GTD the issue of metropolitan consolidation remains on the political agenda, and so the support of international NGOs remains critical to sustaining nascent structures of suburban governance.

Perhaps all of this might prompt us to conclude that the form of suburban governance in Albania is not amenable to prior theoretical claims based on the experiences of suburban spaces and edge cities in North American or other parts of Europe (Hamel and Keil 2015). However, such a conclusion seems premature. At the very least, this case study of one suburban settlement in Albania helps to extend our knowledge of how capitalist property relations continue to generate tensions around the suburban living place. In post-communist Albania, overriding considerations are ongoing claims by migrants and long-standing residents for the right to construct homes and conduct their livelihoods around the suburbs, and how such claims have been supported and sustained by international institutions including the World Bank and NGOs.

NOTES

1 To emphasize Albania's communist legacy, we use the descriptor *post-communist* rather than *post-socialist* in our analysis.
2 Successive plans have been developed for the GDT region by planners and consultants from other European countries and the World Bank. However, the legal status of these plans remains in some doubt.
3 The number of small businesses in Kamza officially registered for tax collection is estimated to have increased from 500 in 2007 to more than 1,200 in 2014 (data from the Kamza Bulletin, 2007–14).
4 The use of the term *migrant* in this context is problematic because of both the mass exodus of many Albanians to neighbouring states such as Greece and Italy during the 1990s and, subsequently, migrations into Albania from neighbouring Kosovo. The "migrants" who have settled in Kamza are mainly internal or "diverted" migrants – people who came to Tirana from rural areas in the northern regions of Albania itself.

REFERENCES

Abitz, Julie. 2006. "Post-Socialist City Development in Tirana." PhD diss., Roskilde Universitet, Speciale Internationale Udviklingsstudier. https://core.ac.uk/download/pdf/12515943.pdf

Aliaj, Besnik, Kedja Lulo, and Genc Myftiu2003. *Tirana: The Challenge of an Urban Development*. Tirana, Albania: Cetis.

Andoni, Doris.2007a. "Legalization of Informal Settlements in Albania." Paper presented at the fifth Regional Vienna Declaration Review Meeting, Stability Pact for South Eastern Europe, Vienna, Austria, 22–3 October.

– 2007b. "The Paradigm of Legalization – A Paradox or the Logic of Development." Paper presented at Workshop on Informal Settlements – Real Estate Market Needs Related to Good Land Administration and Planning, FIG Commission 3 Workshop, Athens, Greece, 28–31 March. http://library.tee.gr/digital/m2267/m2267_andoni.pdf

Andrusz, Gregory, Michael Harloe, and Ivan Szelenyi, eds. 1996. *Cities after Socialism: Urban and Regional Change and Conflict in Post-Socialist Societies.* Oxford, UK: Blackwell.

Childress, Malcolm. 2006. "The Unfinished Business of Land and Property Reform in Albania." In *On Eagle's Wings: The Albanian Economy in Transition,* edited by Dirk J. Bezemer, 115–29. New York, NY: Nova Science.

Co-PLAN (Center for Habitat Development). 2002a. *Making Cities Work! A Culture of Change. 2001 Annual Report.* Tirana, Albania: Co-PLAN. https://issuu.com/co-plan_tirane/docs/2001_annual_report

– 2002b. *Strategic Urban Development Plan of the Municipality of Kamza.* Tirana, Albania: Co-PLAN.

Council of Europe. 2008. *Kamza Development Strategy, 2008–2015.* Tirana, Albania: Council of Europe. http://www.comune.macerata.it/Engine/RAServeFile.php/f/Albania_Kamez_documentazione_citta.pdf

Cox, Kevin R. 2002. *Political Geography: Territory, State, and Society.* Oxford, UK: Blackwell.

Cox, Kevin R., and Andrew E.G. Jonas. 1993. "Urban Development, Collective Consumption and the Politics of Metropolitan Fragmentation." *Political Geography* 12: 8–37. https://doi.org/10.1016/0962-6298(93)90022-y

Cox, Kevin R., and Ron J. Johnston, eds. 1982. *Conflict, Politics and the Urban Scene.* London, UK: Longman.

de Waal, Clarissa. 2004. "Post Property Rights and Wrongs in Albania: An Ethnography of Agrarian Change." *Conservation and Society* 2: 19–50.

Dedja, Taulant, and Fran Brahimi. 2006. "The Dilemma of the Revision of the Administration and Territorial Division in Albania: Obligatory, Voluntary Amalgamation or Inter-Communal Collaboration as a Transitory Solution." Paper presented at the conference on Decentralization Between Regionalism and Federalism in the Stability Pact Countries of the Western Balkans, Tirana, Albania, 15 June.

Felstehausen, Herman. 1999. *Urban Growth and Land Use Changes in Tirana, Albania: With Cases Describing Urban Land Claims* [working paper]. Madison, WI: Land Tenure Center, University of Wisconsin.

Goodwin, Mark, and Joe Painter. 1996. "Local Governance, the Crises of Fordism and the Changing Geographies of Regulation." *Transactions of the Institute of British Geographers* 21, no. 4: 635–48. https://doi.org/10.2307/622391

Hamel, Pierre, and Roger Keil, eds. 2015. *Suburban Governance: A Global View.* Toronto, ON: University of Toronto Press.

Hartkoorn, Adri, ed.2000. *City Made by People*. Tirana, Albania: Co-PLAN (Centre for Habitat Development).

Hirt, Sonia. 2012. *Iron Curtains: Gates, Suburbs, and the Privatization of Space in the Post-Socialist City*. Oxford, UK: Wiley-Blackwell.

Hirt, Sonia, and Atanass Kovachev. 2015. "Suburbia in Three Acts: The East European Story." In *Suburban Governance: A Global View*, edited by Pierre Hamel and Roger Keil, 177–97. Toronto, ON: University of Toronto Press.

Jonas, Andrew E.G. 1991. "Urban Growth Coalitions and Urban Development Policy: Post-War Growth and the Politics of Annexation in Metropolitan Columbus." *Urban Geography* 12: 197–226.

Jonas, Andrew E.G., Eugene McCann, and Mary Thomas. 2015. *Urban Geography: A Critical Introduction*. Oxford, UK: Wiley-Blackwell.

Judt, Tony. 2010. *Postwar: A History of Europe since 1945*. London, UK: Vintage /Random House.

Kelm, K. 2002. "Case Study: Albania." Paper presented at regional workshop on Land Issues in Central and Eastern Europe and the CIS, Budapest, Hungary, 3–6 April.

Logan, John, and Harvey Molotch. 1987. *Urban Fortunes: The Political Economy of Place*. Berkeley and Los Angeles: University of California Press.

Mele, Marcela. 2011. "Transition Stories: Politics of Urban Living Space in Tirana City Region, Albania." PhD diss., University of Hull.

Pickles, John, and Adrian Smith, eds. 1998. *Theorising Transition: The Political Economy of Post-Communist Transformation*. London, UK: Routledge.

Pickles, John, and Adrian Smith. 2005. "Post-Socialism and the Politics of Knowledge Production." In *Politics and Practice in Economic Geography*, edited by A. Tickell et al., 151–62. London, UK: Sage.

Pojani, Doriana. 2013. "From Squatter Settlement to Suburb: The Transformation of Bathore, Albania." *Housing Policy* 28, no. 6: 805–21.https://doi.org/10.1080/02673037.2013.760031

Robinson, Jennifer.2006. *Ordinary Cities: Between Modernity and Development*. London, UK: Routledge.

Sjöberg, Örjan. 1992. "Underurbanization and the Zero Urban Growth Hypothesis: Diverted Migration in Albania." *Geografiska Annaler, Series B: Human Geography* 74, no. 1: 3–19. https://doi.org/10.1080/04353684.1992.11879627

– 2014. "Cases Onto Themselves? Theory and Research on Ex-Socialist Urban Environment." *Geografie* 119: 299–319.

Sluis, Reynt, and FrankWassenberg. 2003. "Urban Developments in Tirana." In *Making Cities Work! The International Conference of ENHR, the European Network of Housing Research*, edited by Besnik Aliaj et al., 15–23. Tirana, Albania: ENHR.

Smith, Adrian. 2007. "Articulating Neoliberalism: Diverse Economies and Everyday Life in Postsocialist Cities." In *Contesting Neoliberalism: Urban Frontiers*, edited by Helga Leitner, Jamie Peck, and Eric Sheppard, 204–22. New York, NY: Guilford.

Stenning, Alison. 2003. "Shaping the Economic Landscapes of Postsocialism? Labour, Workplace and Community in Nowa Huta, Poland." *Antipode* 35, no. 4: 761–80. https://doi.org/10.1046/j.1467-8330.2003.00349.x

Sýkora, Lyděk. 2008. "Revolutionary Change, Evolutionary Adaption and New Path Dependencies: Socialism, Capitalism and Transformations in Urban Spatial Organizations." In *City and Region: Papers in Honour of Jiri Musil*, edited by Wendelin Strubelt and Grzegorz Gorzelak, 283–95. Leverkusen Opladen, Germany: Budrich Unipress and Farmington Hills.

Tsenkova, Sasha. 2010. "Informal Settlements in Post-Communist Cities: Diversity Factors and Patterns." *Urbani izziv* 21, no. 2: 73–84. https://doi.org/10.5379/urbani-izziv-en-2010-21-02-001

World Bank. 2007a. *Albania Urban Sector Review (English)*. Washington, DC: The World Bank. http://documents.worldbank.org/curated/en/125501467992047283/Albania-urban-sector-review

– 2007b. "World Bank Responses to the Problem of Informal Development: Current Projects and Future Action." Paper presented at workshop on Informal Settlements – Real Estate Market Needs Related to Good Land Administration and Planning, FIG Commission 3 Workshop, Athens, Greece, 28–31 March.

Zickel, Raymond E., and R. Walter Iwaskiw, eds. 1994. *Albania Country Study*. 2nd ed. Washington, DC: US Library of Congress, Department of the Army.

12 Urban Natures in Managua, Nicaragua

LAURA SHILLINGTON

In an editorial published at the end of the 1980s in Nicaragua's main academic journal, *Revista Envio,* several social scientists argued that Managua had become an "odd city" because of two main forces "at work on the city: earthquakes and capitalism" (Equipo Nitlápan-Envío 1989, my translation). This chapter explores not only the imprints of capitalism and earthquakes in the city, but also discourses of modernism and socialism that equally continue to leave their marks. It shows how Managua has developed through socio-natural relations. More specifically, the chapter looks at two prominent socio-natural relations that define Managua: lakes and earthquakes.

Managua, the capital of Nicaragua, is located on the southern shores of Lake Xolotlán (or Lake Managua) and has within its political boundaries four volcanic crater lakes called *lagunas* – Asososca, Nejapa, Tiscapa, and Acahualinca (figure 12.1). The history, culture, and climate of Managua are intimately linked to Lake Xolotlán and the lagunas. Throughout the twentieth century, the lagunas and lake have been sources of drinking water, food, income, recreational areas, and tourist attractions as well as receptacles for garbage, raw untreated sewage, and industrial chemicals (Santos López 1982). Earthquakes are also key players: There are hundreds of active seismic fault lines that zigzag across the city, generating small tremors almost weekly. In the twentieth century, Managua was radically altered twice by major earthquakes. The first, in 1931, ravaged a large part of the city's core. The second, in 1972, was more devastating and destroyed more than three quarters of the city. The destruction caused by both earthquakes led to an inclination towards low-density, low-level urban buildings; very few structures over three storeys were built until the mid- to late 1990s. Unlike many cities in Latin America (and globally), Managua's skyline remains largely horizontal, with a canopy of trees towering over most structures.

But while the earthquakes have been powerful forces in shaping Managua's landscape, they also played a part in shifting the political and economic landscape (Rodgers 2011). Nature – in particular earthquakes

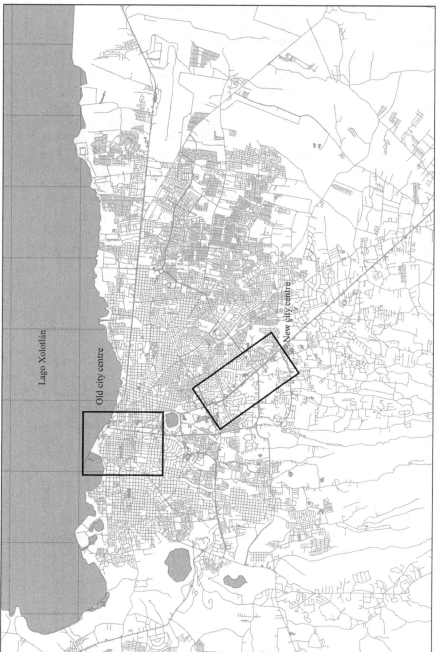

12.1 Map of Managua showing the four *lagunas* along with '"new" and old centres. Map data © OpenStreetMap contributors, and available from https://www.openstreetmap.org.

and lakes – has shaped Managua and, simultaneously, the political and economic transformations have created particular urban natures. This chapter considers these socio-natural transformations within the contexts of the Sandinista socialist government in the 1980s, the neoliberal governments of the 1990s and early 2000s, and finally the present incarnation of the Sandinista administration.

The chapter begins by situating Managua with a discussion of the periods following the 1931 and 1972 earthquakes up until the Sandinista revolution in 1979. The following sections examine the role of nature in urban development during the three periods mentioned above. Within all these eras, different forms of urban nature(s) were privileged and produced in various ways. This chapter explores these diverse natures and the emerging political ecologies. The chapter uses an urban political ecology approach, which understands the city and urbanization as outcomes of both socio-economic and natural processes – socio-natural processes (Heynen 2014) – to understand how Managua has evolved as a city.

Modernizing Managua: From Somocismo to Sandinsimo

In 1931, the centre of Managua was destroyed by a 6.0 magnitude earthquake. What the earthquake did not level was later damaged by numerous fires. The city became a blank slate for the political desires of one powerful Managuan family, the Somozas. In 1936 Anastasio Somoza García seized power and established an authoritarian dictatorship that lasted four decades, with his sons Luis Somoza Debayle and Anastasio Somoza Debayle succeeding him. When Somoza García took power shortly after the earthquake, urban modernization became his main priority.[1] Reconstruction efforts following the 1931 earthquake attempted to replace older, adobe colonial buildings with concrete and steel. Somoza's aim was to construct a downtown core similar to industrial cities like Chicago and Detroit. Indeed, he spent much time in the United States (especially Miami) and was determined to "westernize" Managua and make it comparable to US cities. Downtown Managua after 1931 (but prior to 1972) was made up of Spanish colonial buildings intermingled with modern buildings such as high-rises.

As part of what has been called the "Managua Strategy" (Higgins 1990), the Somoza government established the *Oficina Nacional de Urbanismo* (National Office of Urbanism) in the early 1950s, and in 1954 the office published the first planning documents for the city. The plans drew heavily on urban planning in the US. The city was zoned following a monocentric structure, where Managua's centre was the main focal point and all other zones radiated out from this centre (Godoy 1988).[2] These plans had a particular focus on road infrastructure. They also improved water and sewage infrastructure and electricity in "authorized" settlements, which were those settled under the market-based real estate system and public housing (upper- and middle-class areas of the city; Morales 1992, 1998). Urban services for illegal subdivisions (*repartos ilegales*) or *tomas de tierra* (literally, the taking of land) were in many cases never developed.[3]

The lake and lagunas factored into Somoza's modern Managua in two ways. The first was as receptacles for waste. The first sewerage system was constructed in 1927 and directed all storm- and wastewater directly into the lake without treatment (IRENA 1982). Under Somoza this continued, with any new sewer pipes flowing straight into the lake. Much of the contamination in the storm- and grey water derives from toxic chemicals leaching from solid waste deposited in informal garbage dumps throughout the city. The main municipal waste site was created under the Somoza regime in 1943 and located between the shores of Lake Xolotlán and Laguna Acahualinca on the western edge of Managua (Arguello Herrera 1982). While there were numerous informal dumping sites scattered throughout the city, efforts were made under the Somoza government to centralize garbage into this main site – as was done in American cities at the time. This garbage dump, known as *la Chureca*, slowly grew over three decades with the increasing production of domestic, commercial, and industrial waste. As it grew, not only did toxins leach into both bodies of water, but solid waste also entered into the Laguna Acahualinca. The Chureca dump would become a key site of social and environmental struggles in the later decades of the twentieth century (Hartmann 2012).

Not all waste was directed to the Chureca dump. Improvements in infrastructure for industry in the eastern area of the city were prioritized to help attract foreign investment (Rodgers 2008). Locating industry along the lakeshore was strategic because the lake provided a cheap source of water while serving as a repository for waste. The lack of infrastructure to properly deal with industrial waste and the absence of environmental regulations resulted in unrestrained dumping of effluents (including heavy metals such as mercury) into the lake, making it one of the most polluted lakes in the world (Lacayo et. al. 1991; McCrary, Castro, and McKaye 2006). Consequently, the marginal lands and the inhabitants' living conditions along the lake were worsened. Many informal settlements (or tomas de tierra) had emerged along the lake over the decades following the earthquake. The inhabitants depended on the lake for drinking water and livelihoods (mainly fishing).

As the water became more polluted their drinking water and food were also contaminated. Despite such costs, the industrialization was critical for the modernization of Managua and necessary for the Somocistas to increase their economic power in the country (Rodgers 2008).

The second way that the Somoza regime understood the lake and lagunas was as resources for exploitation. Lake Xolotlán was a source of cheap water for industries. The lake had not been used as a source of potable water since 1925, when the city started to exploit Laguna Asososca, which has served as the main source of potable water along with a series of aquifers located in various parts of the city (Alcaldía de Managua and MIVAH 1988; Bethune et al. 1996; IRENA 1982). The lake also served as a recreational site, but primarily in the early years of the dictatorship. The recreation encouraged by the Somoza-run city government was aimed at wealthy Nicaraguans and foreigners. The recreation activities had little to do with the lake itself, but involved clubs and casinos. In the late 1940s a dock was built from the old pre-1931 earthquake *malecón* (lake-front promenade) to house what became Managua's famous Casino de la Playa, also known as Casino Casablanca. Very little was invested into rebuilding the old malecón for public use or cleaning the lakeshore, which by 1969 the government supposedly declared uninhabitable because of contamination (Silva 2009). The lagunas were also not considered possible spaces for public recreation. Laguna Tiscapa, overlooking the historic centre, and Lake Xolotlán were the location of the presidential residence and contained a small park for the public but none of the other lagunas were exploited for recreation.

The modernization efforts under the Somoza regime produced a city resembling cities in North America in many ways, with distinct commercial, residential, and industrial zones. Wall (1996) argues that Managua resembles Chicago more than it does other Latin American cities, primarily because of the lake and the necessity of developing in one direction. Higgins (1990, 383) describes how the spatial reorganization under Somoza "included a land use policy of suburbanization," which sought to emulate the newer suburban developments occurring at that time in the United States. In the early 1970s, the suburbanization of Managua intensified. Both the Luis Somoza Debayle and Somoza-controlled municipal governments still concentrated much of their efforts in the centre of the city, attempting to create a more modern downtown. In the centre large paved boulevards and avenues, brick sidewalks, and street signs were constructed, all of which are virtually non-existent in the city today.

The modernization of Managua shifted dramatically in December 1972 when the city was levelled by two successive earthquakes: the first one measuring 6.3 and the second (less than an hour later) 6.4 on the Richter scale, both with epicentres in the city's centre. The earthquakes demolished the entire city centre. It is estimated that over 80 per cent of the city's housing was destroyed, one quarter of its heavy industry, and 90 per cent of its commercial capacity (Chávez 1987; Higgins 1990). The death count was reported to be close to 100,000, reducing Managua's population from 425,000 to less than 350,000 (Higgins 1990). The majority of older colonial buildings, weakened by previous earthquakes, were completely destroyed. Not only were the low-lying colonial structures destroyed but also the more modern high-rises. The high-rises left standing were bulldozed not long after reconstruction efforts began because their structure had been damaged beyond repair.

Consequently, the landscape of Managua was completely altered – structurally, economically, politically, and socially. Writing in 1987, Doreen Massey observed that the earthquake changed the form of Managua beyond recognition: "The centre was left empty, the roads criss-crossing the open space in an eerie reminder of the past, and weeds and wild flowers gradually took over what once had been the central blocks of the capital" (Massey 1987, 10).

While the national government received millions in international economic aid (both multilateral and bilateral) for reconstruction efforts, attempts at rebuilding Managua were inefficient and plagued by corruption. Unlike the relatively quick reconstruction following the 1931 earthquake, Managua is still trying to recover more than forty years later. The foreign financial aid intended for reconstruction was controlled almost exclusively by the national government, or more correctly, as Higgins (1990) clarifies, by the Somocistas. President Anastasio Somoza Debayle made himself head of the reconstruction committee. He could thus direct all reconstruction contracts to his own construction companies and those of Somocistas, thereby appropriating most of the international aid.[4] Housing in particular became a major source of income for the Somocistas since the majority of foreign aid was earmarked for rebuilding housing and urban infrastructure. To ensure that his economic wealth and power would increase, Somoza made sure that a large majority of housing and urban infrastructure projects were realized on land he and his close friends owned. Much of this land lay at the periphery of the city, enabling full implementation of the suburbanization policies established prior to the earthquakes (Inzulza Contardo and López Irías 2014; Revels 2014; Wall 1996). As several Nicaraguan social scientists note, the city turned into a "Third World Los Angeles laid out for the motorcar with suburbs spread about shopping malls. Yet in Managua there [were] few cars" (Equipo Nitlápan-Envío

1989, my translation). The damaged areas of the city were completely ignored. The consequence was, as Wall (1996, 48) notes, a "massive internal restructuring of the city"; so much so that the 1972 earthquake and failed reconstruction efforts remain physically visible in the present landscape. Nowhere is this more evident than in the still relatively empty "old" city centre. The centre of Managua is only now being rebuilt, and there are still large vacant areas and remnants of destroyed buildings.

At the same time that the physical form of Managua was altered, the Somocistas' power also shifted. Ironically, while the 1972 earthquake increased Somoza's wealth and economic power, it decreased his political power (see Galeano 2000). People's general dissatisfaction with the dictatorship that had begun in the 1960s increased significantly in the years following the earthquake. Discontent with the successive Somoza government was felt across classes, to the extent that even the upper class (those not part of the Somocistas) joined forces with the *Frente Sandinista Liberación Nacional* (FSLN). Only seven years after the earthquake the FSLN and their supporters ousted dictator Anastasio Somoza Debayle and the Somocistas. The change of power in 1979 resulted in a significant alteration of Nicaragua's economic, social, and physical landscape, and Managua as the capital saw many of these changes.

¡Viva la revolución! Sandinismo and Socialist Urbanism

The FSLN was founded as a political organization in 1961, with its main centres at universities in Managua and León. Inspired by the revolution in Algeria, the FSLN aimed to overthrow the Somoza dictatorship and bring democracy to Nicaragua. The name *Sandinista* pays tribute to Augusto César Sandino, who in the early twentieth century (1920s) led the rebellion against the US occupation of Nicaragua. Many leaders of the FSLN, more commonly known as Sandinistas, received military training in Cuba and later returned to train civilians. By early 1979 they had a large army and held power throughout Nicaragua, with the exception of Managua. On 19 July 1979 the Sandinistas ousted Somoza from power, defeating the National Guard and taking over the *Palacio Nacional*. The FSLN sought to establish a socialist democracy in Nicaragua.

When the FSLN took power in 1979, Nicaragua was a country with a severely damaged economic and physical landscape. At the time, Managua did not have any comprehensive long-term planning agenda or process (Chávez 1987). Given that reconstruction efforts were focused on benefitting primarily Somoza and urban elites, the Sandinistas inherited an exceedingly socio-spatially segregated city. More than a quarter of the land was owned by less than 2 per cent of the population, and the majority of wealthy Managuans lived in the new suburban settlements developed by Somocistas. More than half of the city's residents lived in illegal settlements and approximately 50 per cent of usable urban land was vacant (Alcaldía de Managua and MIVAH 1988; Chávez 1987). Paradoxically, while the majority of these illegal settlements and slums had no urban infrastructure (no sewage, potable water, electricity, or proper roads), over half of the unused land (especially the areas damaged by the earthquake and not resettled and those owned by Somoza or wealthy elites) had complete electricity, water, and sewage networks (Higgins 1990). Some of this was in part because of international aid that Somoza had diverted to install urban infrastructure on his own land. His land and that of most of the other wealthy elites was appropriated by the Sandinistas and, along with the vacant land, became key to their urban reform.

Driven at the onset by discontent with the unequal land distribution (both rural and urban), the Sandinista revolutionary government made rural and urban land reform the core issue of the new government. Land reform was part of a larger regional strategy that envisioned a more balanced population density between the Pacific, North-Central, and Atlantic areas. In this national spatial strategy of equalizing rural-urban development, urban reform sought to direct development away from Managua and the Pacific region and towards less-populated urban and rural areas (Chávez 1987; Wall 1993). Under this strategy, the government created a policy to make Managua "less attractive to emigrate to and rural areas and regional centres correspondingly more attractive" (Equipo Nitlápan-Envío 1989, my translation). A main component of this policy was the legalization and titling of land along with the promise to provide urban benefits (water, electricity, etc.) to all residents (Lungo and Morales 2000; Wall 1993). Although the government attempted to discourage migration into the city by enticing people to stay in rural areas through agrarian reforms, migrants continued to move into Managua. As Chávez (1987, 288) points out, the redistribution of "the benefits of urbanization among the poor majority of the population" was considered unachievable in rural areas, which thus encouraged continued migration. Despite the national strategy to take focus away from Managua, new urban developments materialized throughout the 1980s.

In addition to land reforms, the national government established the *Junta de Reconstrucción de Managua* (Council for Reconstructing Managua, JRM), which was charged with reconstructing Managua (IRENA 1982). They succeeded in clearing most of the rubble (from the earthquake and insurrection) from the city centre and established a large public park, several small green areas, and public housing units (Lungo and Morales 2000). Because of the many active seismic fault lines zigzagging the city centre, there was little desire to fully reconstruct. Therefore, some of the vacant land in the old centre (as well as throughout the city) was converted into agricultural land – approximately 9 per cent of the land within the city limits (Equipo Nitlápan-Envío 1989).[5] The use of urban land to grow crops for a rapidly increasing urban population was critical, and because of the risk of earthquakes and flooding in the old centre agriculture became one of the only ways to make the city centre "productive" and useful.

The JRM along with others in the Sandinista administration at the city and national levels were very concerned about the contamination of Lake Xolotlán. In November and December of 1982 an international workshop on recuperating the lake was held jointly by JRM and the *Instituto Nicaragüense de Recursos Naturales y del Ambiente* (Nicaraguan Institute of Natural Resources and the Environment; IRENA 1982). Aiming to create a national action plan for the lake, the workshop was organized around five central themes relating to problems and possibilities of Lake Xolotlán: (1) water quality; (2) the potential of hydropower; (3) potential risks and hazards; (4) development and management of the watershed; and (5) city-lake relations. The workshop recommended building sewage and water treatment plants, relocating the Chureca garbage dump to a larger site (the dump had expanded exponentially primarily because the debris from the earthquake was deposited here), reconstructing the malecón, creating public parks on the lakeshore, ensuring strict environmental regulations for industry, and relocating industry away from the lake. The final workshop document explicitly equates the reclamation of the lake with ridding the country of capitalist imperialism and the vestiges of the dictatorship. Indeed, the workshop's final presentation was directed at the "Commandant of the Revolution" (Daniel Ortega) and ended by stating that "the salvation of Lake Managua is a problem of national dignity; the salvation of Lake Managua is a problem of free the homeland or die" (IRENA 1982 131).

In spite of the detailed recommendations set out by the workshop, very few of them came to fruition. When the Sandinistas took power, anti-Sandinista forces (contras) supported both economically and militarily (especially in training and weapon supplies) by the United States began an offensive to topple the Sandinista government. In addition to backing the contra forces, the United States also imposed economic sanctions on Nicaragua almost as soon as the Sandinistas took power. These economic sanctions and the costs of fighting the contras prevented the Sandinista government from fully implementing their economic and social policies, including the reclamation of Lake Xolot- lán and rebuilding Managua. The intensity of the contra wars and the economic repercussions prompted the Sandinistas to call an official state of emergency. The national elections in 1984 were held under the state of emergency and the Sandinistas were voted into power with just over 66 per cent of the votes. The following national election in 1990 saw the defeat of the Sandinistas (with 40 per cent of votes) and the entry of the US-supported government of Victoria Chomorro of the coalition party *Unión Nacional Opositora* (UNO), with just over 50 per cent of votes.

The Miamization of Managua: Neoliberal City and *La Ciudad Popular*

Under the new UNO conservative government, the desire to make Ma- nagua a modern city was once again the focus of the national govern- ment. Chamorro's government was the first of three liberal conservative parties that held power in Nicaragua from 1990 until 2006 (Alemán and Bolaños followed Chamorro). After the 1990 elections, the socialist ide- als and policies of the Sandinista government were quickly replaced with capitalist concerns about the global market. In many ways, Somo- za's ambition to "modernize" Managua was re-invigorated, initiated strongly by the return of Nicaraguans who had fled to the United States during the insurrection and in the early 1980s (Babb 1999). Returning Nicaraguans and their US-born children (many of them teenagers in the early to mid-1990s) resettled primarily in the capital, and brought with them particular notions of urbanity based on their experiences of living in US cities (mainly Miami), in addition to considerable amounts of money to invest in Nicaragua. These repatriots, or "Miami boys" as they are called, have sought to "recreate their cherished Miami social and cultural scene" (Whisnant 1995, 448).[6]

The cityscape of Managua was transformed to resemble Miami. The Miami boys and their US capital imported North-American-style urban (or rather suburban) elements: international franchises (e.g., Pizza Hut and Subway), luxury restaurants and hotels, exclusive North-American-style supermarkets, three luxury shopping malls, private English schools (teaching American curriculum to facilitate entrance into US universities), and gated housing communities. To accompany these new developments, the city expanded their already extensive (and under-maintained) road network: several new four-lane thoroughfares were constructed in the early 2000s, including the ex- pansion and improvement of the Masaya highway, along which the majority of new urban developments are located (once again ignoring the old city centre; see figure 12.1). Urban elites created a new city centre along this highway, although most Managuans continued to refer to the old centre as the city centre.

These developments during the 1990s and 2000s once again favoured urban elites. As it did in the Somoza era, the real estate sector boomed. Babb (1999, 47) points out that "neoliberal policies of structural adjustment mandated by the International Monetary Fund and the World Bank" determined social spending and infrastructure development. President Alemán, in particular, sought to implement a neoliberal development model. He had previously served as mayor of Managua under Chamorro and, as Babb (1999) comments, built his reputation on a project to modernize Managua. One of Alemán's projects as mayor was rebuilding the malecón, and then later as president he built a new presidential palace (Rodgers 2008). However, the majority of development was concentrated in the south of the city, which resulted in the centre – and the malecón – falling into disuse once again.

What differed from the Somoza modernization era was an obligation to recognize and address environmental concerns. Discourses of sustainability framed much of the development in Nicaragua (and elsewhere) during the 1990s. Environmental problems could not be ignored within the global context of sustainable development, especially for securing large funds from the World Bank and International Monetary Fund. The liberal conservative government engaged in projects that, at least on the surface, attended to Managua's environmental problems. The central concern was on the contamination of Lake Xolotlán coupled with the lack of a city-wide sewer system and sewage treatment plants. The sewage treatment plants were a central focus in large part because they fit in easily with neoliberal processes of privatization, commodification, and deregulation. Alemán successfully privatized the national electricity company to Spanish multinational Unión Fenosa, and attempted do the same with water and sanitation.

The government of Alemán created legislation in 1998 (Law 297, *la Ley General de Servicios de Agua Potable y Alcantarillado Sanitario*) to restructure the national distribution system of water and sanitation. This restructuring was financed by an Inter-American Development Bank (*Banco Interamericano de Desarrolo*, BID) loan. As a result, the *Instituto Nicaragüense de Acueductos y Alcantarillados* (Nicaraguan Water Supply and Sewerage Institute, INAA), which had been responsible for setting policies around water and sewerage, as well as the regulation and the distribution of services at a national level, was separated into three organizations. The INAA, maintained as a government organization, had its main responsibilities reduced to just the regulation of the system. This includes protecting consumers' rights to access, establishing procedures and norms around the distribution and pricing of water and sewer services, and the administration of concessions. A second organization, the *Comisión Nacional de Agua Potable y Alcantarillado Sanitario* (National Commission on Potable Water and Sewer, CONAPAS) was created to oversee the formulation of policy and planning (INIFOM 2002; INAA and BID 1998). Finally, the system's operation was given to the newly created public utility company, the *Empresa Nicaragüense de Acueductos y Alcantarillados* (Nicaragua Water and Sewer Company, ENACAL). This modernization program is what Castree (2008, 142) has described as a "state-led attempt to run ... public services along private sector lines as 'efficient' and 'competitive' businesses." Indeed, ENACAL was created with the aim of contracting out management and service delivery to a private international operator and transforming it into a publicly traded company (Equipo Nitlápan-Envío 2006). Although initiated in 1999, the transformation of ENACAL from public to private was met with large-scale resistance and did not occur.

In addition to neoliberalizing the water sector, the Chamorro, Alemán, and Bolaños governments (1996–2006) also laid the groundwork for the construction of Managua's wastewater treatment plant. Numerous studies and reports were produced in collaboration with the BID (e.g., INAA and BID 1998). Work on the plant began shortly afterwards with funding from the BID and Germany, but the plant experienced numerous problems and was not operational until 2009. It was the Sandinista government of Daniel Ortega that inaugurated the plant (Gómez 2009), which is managed by the public company ENACAL. While the plant represents a step forward in improving the conditions of the lake, it does not extract and purify the lake water; it only treats incoming water and waste from the city. In this regard, the water in Lake Xolotlán remains highly contaminated – even breathing air near the shore is considered dangerous (Álvarez 2009). This complicates the current government's desires to revitalize and expand the malecón to create new cultural-natural relations with the lake.

Neoliberal Socialism? Sandinismo of the New Millenium

The lakeshore, malecón, and historic centre have become key sites of change under the current Sandinista government, which has been in power since 2007 under the leadership of Daniel Ortega. Lake Xolotlán has become a central player, drawing on the discourses employed in the 1980s of the lake as integral to Managuan and Nicaraguan identities.

Managua's historic centre is being linked in several ways to the lake on which it sits. The old historic centre has been converted into public spaces, including a large park – Parque Luis Alfonso Velásquez Flores – featuring over six different play structures for children, basketball courts, and baseball diamonds. It is located just off Avenida Bolivar, which leads from Laguna Tiscapa to the lakefront and is used on special occasions as an outdoor public gallery that displays, on rotation, large works of art. This is not particularly good for traffic, but it has attracted Managuans and tourists into the centre, and to the new port and malecón.

Puerto Salvador Allende was constructed in 2008 and now serves as an *embarcadero* for the revived ferries that take tourists on lake excursions. It is one of two designated "tourist" ports in the city (the other is San Juan del Sur on the south Pacific coast). Beside the lakeshore port sits the revived malecón with its restaurants, bars, and a playground for children. The malecón hosts events such as concerts and sporting events (boxing) weekly. Just east of the port and malecón is Paseo Xolotlán. Called a family park, Paseo Xolotlán was built to celebrate thirty-five years of the Revolution. It consists of a children's play area (the focus on playgrounds over the past several years in Managua is notable) as well as a model of the "glamorous Avenida Roosevelt" as it looked prior to the 1972 earthquake (Peréz 2014).

What is interesting about the developments along Lake Xolotlán is the mixture of socialist desire for collective identity and public space with nostalgia for a non-socialist-era (and elite) urban landscapes and neoliberal policies. Visitors have to pay 40 Córdobas (US$1.50) to enjoy the lakeshore (port and malecón), which is beyond the reach of many Nicaraguans. However, Paseo Xolotlán and the other parks in the centre are free, unless the visitor needs to park. The malecón in many ways is understood more as a tourist attraction and revenue generator for the city. Yet there is a strong desire on the part of the city and national government (both Sandinista) to create in Nicaraguans an attachment to the lake both as "nature" and resource. Recently, the city began a project to plant trees, plants, and grasses along the lakeshore to help naturally filter the water and stablize the shore. The current Sandinista government has carried over IRENA's 1980s concern with cleaning the lake water for recreational use and also as a source of geothermal power (IRENA 1982). In May 2015, the government successfully secured funding from the German government to expand the wastewater treatment plant, which would include capturing biogas for energy.

Cleaning the water has again become a symbol of socialist desire to decrease imperialism in the country. However, while the project as it was framed in the 1980s was nationalistic (see IRENA 1982), it is no longer just national. Instead, the new Ortega administration has melded nationalist with neoliberal desires. As Rodgers (2008, 120) points out, the administration has "mixed anti-imperialist rhetoric and *rapprochement* with Chávez's Venezuela's trouble-free negotiations with the IMF [and] big business." At the same time, there is a deep nostalgia for the Managua of the 1960s – the bustling, energetic, and more important, clean and ordered city – a nostalgia that draws chiefly on the *elite* urban landscape of the 1960s. In addition, there is also a strong desire in contemporary discourses to create Managua in the likeness of Miami, and perhaps a little bit like the more sustainable European cities.

Conclusion

Managua has been shaped by earthquakes and lakes as well as political and economic transformations. Through conservative dictatorship, socialist revolutionary, neoliberal, and (pseudo-) socialist governments, Managua has had a changing relationship with its natures. Lake Xolotlán has been central to how Managua relates to and understands urban socio-natures. The lake is connected to urban development (through sewers, storm canals, and garbage), but in many political regimes, the lake has been forgotten or only noticed in times of crisis. The Sandinista government during the 1980s saw the lake as intimately connected with revolutionary goals. The postsocialist, neoliberal era that followed the defeat of the Sandinistas from 1990 to 2006 governed nature through neoliberal processes of privatization, commodification, and deregulation, guided by global discourses of sustainable development. The current Sandinista government has drawn on a complex combination of socialist ideals, nostalgic desires for the modern Managua of the 1950s and 1960s, and neoliberal processes, which has created an interesting and distinctive arrangement of urban nature and political ecologies.

NOTES

1 Rodgers (2008, 2011) explores in depth Managua's urban development since the 1930s and in particular over the past two decades.
2 This type of planning followed much of what was put forward by Chicago School urbanists such as Ernest Burgess and Robert E. Park.
3 Interview with Ninnette Morales Ortega, Director of El Centro HABITAR, Managua, Nicaragua, 19 June 2006. See also Morales (1992).

4 See Higgins (1990) for a more detailed discussion on Somoza's control of reconstruction efforts in Managua.
5 Interview with Guillermo Martínez, agronomist at Fundación Nicaraguense Prodesarrollo Comunitario Integral (FUNDECI), Managua, Nicaragua, 3 August 2005.
6 The label "Miami Boys" has recently been appropriated by a powerful *pandilla* (or gang) in Managua (García 2006).

REFERENCES

Alcaldía de Managua and Ministerio de la Vivienda y Asentamientos Humanos (MIVAH). 1988. *Esquema de desarrollo urbano de Managua 1987–2020*. Managua, Nicaragua: Alcaldía de Managua.

Álvarez, Gustavo. 2009. "Aguas del Xolotlán ni para las plantas." *El Nuevo Diario*, 1 March, n.d.

Arguello Herrera, Ottoniel 1982. "Problemática de la calidad del agua del Lago de Managua." In *Taller Internacional de Salvamento y Aprovechamiento Integral del Lago de Managua*, 17–54. Managua, Nicaragua: Instituto Nicaragüense de Recursos Naturales y del Ambiente (IRENA).

Babb, Florence. 1999. "Managua Is Nicaragua: The Making of a Neoliberal City." *City and Society* 11: 27–48.

Bethune, David N., R.N. Farvolden, M. Cathryn Ryan, and Andres Lopez Guzman. 1996. "Industrial Contamination of a Municipal Water-Supply Lake by Induced Reversal of Ground-Water Flow, Managua, Nicaragua." *Groundwater* 34: 699–708.

Castree, Noel. 2008. "Neoliberalising Nature: The Logics of Deregulation and Reregulation." *Environment and Planning A* 40: 131–52.

Chávez, Robert. 1987. "Urban Planning in Nicaragua: The First Five Years." *Latin American Perspectives* 14: 226–36.

Empresa Portuaria Nacional (EPN). N.d. Photos of Puerto Salvador Allende. Accessed 15 May 2015 at http://www.epn.com.ni/index .php?option=com_k2&view=item&layout=item&id=22&Itemid=463

Equipo Nitlápan-Envío. 1989. "There Is Nowhere Else Quite Like Managua." *Revista Envío* 91 (February). http://www.envio.org.ni/articulo/2767

Galeano, Marcia Traña. 2000. *Apuntes sobre la historia de Managua*. Managua, Nicaragua: Aldilá Editor.

Garcia, Ernesto. 2006. "Sueltan a 'Los Miami Boys.'" *El Nuevo Diario*. 2 February http://www.elnuevodiario.com.ni/ 2006/02/06/sucesos/12035.

Godoy, Jorge. 1988. "La transformación territorial de Managua entre 1950 y 1979." In *La estructuración de las capitales centroamericanas*, edited by R. Fernández Vásquez and M. Lungo Unclés, 319–39. San José, Costa Rica: EDUCA.

Gómez, Omar. 2009. "Inauguran la Planta de Tratamiento de Aguas." *El Nuevo Diario*, 20 February.

Hartmann, Chris. 2012. "Uneven Urban Space: Accessing Trash in Managua, Nicaragua." *Journal of Latin American Geography* 11, no. 1: 143–63. https://doi.org/10.1353/lag.2012.0003

Heynen, Nik. 2014. "Urban Political Ecology I: The Urban Century." *Progress in Human Geography* 38, no. 4: 598–604. https://doi.org/10.1177/0309132513500443

Higgins, Bryan. 1990. "The Place of Housing Programs and Class Relations in Latin American Cities: The Development of Managua before 1980." *Economic Geography* 66, no. 4: 378–88. https://doi.org/10.2307/143971

Instituto Nicaragüense de Acueductos y Alcantarillados (INAA) and Banco Interamericano de Desarrollo (BID). 1998. "Proyecto Para la Actualización de Plan Maestro de Alcancatarillado Sanitario para la Ciudad de Managua: Segundo Informe Intermedio." Managua, Nicaragua: Author.

Instituto Nicaragüense de Fomento Municipal (INIFOM). 2002. "Ficha Municipal: Managua." Accesed 4 January 2007 at http://www.inifom.gob.ni/ArchivosPDF/Managua2.pdf

Instituto Nicaragüense de Recursos Naturales y del Ambiente (IRENA). 1982. *Taller Internacional de Salvamento y Aprovechamiento Integral del Lago de Managua*. Managua, Nicaragua: Author.

Inzulza Contardo, Jorge, and Néstor López Irías. 2014. "Gentrificación de escala intermedia global en Latinoamérica: El caso de la reconstrucción de Managua, Nicaragua 1972–2014." *Revista de Urbanismo* 16, no. 31: 56–75. https://doi.org/10.5354/0717-5051.2014.33274

Lacayo, M., A. Cruz, J. Lacayo, and I. Fomsgaard. 1991. "Mercury Contamination in Lake Xolotlán (Managua)." *Hydrobiological Bulletin* 25, no. 2: 173–6. https://doi.org/10.1007/bf02291251

Lungo, Mario, with Ninette Morales. 2000. "La tierra urbana pública en Managua durante el gobierno Sandinista." In *La tierra urbana*, edited by M. Lungo, 113–36. San Salvador, El Salvador: Universidad Centroamericana José Simeón Cañas.

Massey, Doreen. 1987. *Nicaragua*. London: Open University Press.

McCrary, Jeffrey K., Mark Castro, and Kenneth R. McKaye. 2006. "Mercury in Fish from Two Nicaraguan Lakes: A Recommendation for Increased Monitoring of Fish for International Commerce." *Environmental Pollution* 141, no. 3: 513–18. https://doi.org/10.1016/j.envpol.2005.08.062

Morales Ortega, Ninette. 1992. "El habitat popular nicaragüense: Las tomas de terreno urbano." *Documento de Estudio* 10. Managua, Nicaragua: FUNDASAL.

– 1998. "La regularización de los asentamientos ilegales en Managua." In *Economía y desarrollo urbano en Centroamérica*, edited by M. Lungo and M.

Polése, 225–68. San José, Costa Rica: Facultad Latinoamericana de Ciencias Sociales (FLACSO).

Revels, Craig. 2014 "Placing Managua: A Landscape Narrative in Post-Earthquake Nicaragua." *Journal of Cultural Geography* 31: 81–105. https://doi.org/10.1080/08873631.2013.873300

Rodgers, Dennis. 2008. "A Symptom Called Managua." *New Left Review* 49: 103–20.

– 2011. "An Illness Called Managua: 'Extraordinary' Urbanization and 'Mal-development' in Nicaragua." In *Urban Theory beyond the West: A World of Cities*, edited by T. Edensor and M. Jayne, 121–36. New York, NY: Routledge.

Santos López, Samuel. 1982. "Relaciones Lago – Ciudad Managua." In *Taller Internacional de Salvamento y Aprovechamiento Integral del Lago de Managua*, 119–24. Managua, Nicaragua: Instituto Nicaragüense de Recursos Naturales y del Ambiente (IRENA).

Silva, Jose. 2009. "Ambiente-Nicaragua: Managua quiere mirarse otra vez en el lago Xolotlán." *Inter Press Service*, 25 February. http://www.ipsnoticias.net/2009/02/ambiente-nicaragua-managua-quiere-mirarse-otra-vez-en-el-lago-xolotlan/

Wall, David. 1993. "Spatial Inequalities in Sandinista Nicaragua." *Geographical Review* 83, no. 1: 1–13. https://doi.org/10.2307/215376

– 1996. "City Profile Managua." *Cities* 13, no. 1: 45–52. https://doi.org/10.1016/0264-2751(95)00093-3

Whisnant, David E. 1995. *Rascally Signs in Sacred Places: The Politics of Culture in Nicaragua*. Chapel Hill: The University of North Carolina Press.

13 The Reshaping of Post-Socialist Ho Chi Minh City: Leisure Practices and Social Control

MARIE GIBERT AND EMMANUELLE PEYVEL

This chapter explores the making of post-socialist Ho Chi Minh City through a socio-spatial analysis of young urban professionals' leisure practices. This angle allows reading the socialist legacy lingering in contemporary Ho Chi Minh City through four dimensions: materiality and architecture, governance, political ideology, and dwellers' practices and representations.

If the name "Ho Chi Minh City" itself is evidence of the socialist period of the Southern Vietnamese metropolis, its socialist regime lasted only a decade, from 1975 to 1986. Thus, Ho Chi Minh City was only marginally affected by socialist urban planning. However, we envision it as post-socialist not only because its contemporary and globalized economic period comes chronologically after a collectivist period (*bao cấp*) but also because both material and immaterial legacies of this period are still prevalent in the contemporary city's production and mutations – in more or less visible and direct ways – from political, economic, social, and cultural perspectives.

The first type of socialist legacy is material and architectural, since leisure has always been at the core of urban projects. Most sites dedicated to leisure created in Ho Chi Minh City during the period of the centrally planned economy are still in use today: they left a longstanding footprint in the city, as shown for instance by the many cultural houses and leisure clubs. The second legacy relates to aspects of governance: while private companies increased rapidly in the leisure sector after 1986, the public sector has remained a key stakeholder. The role of the powerful holding Saigon Tourism, which belongs to the city's popular committee, illustrates this dimension. The third type of legacy is political and ideological, even though the understanding of leisure and its social functions has undergone a semantic renewal since the opening reforms of 1986 (*Đổi mới*).[1] Leisure was first used to promote the modernist idea of universal progress and a heightened degree of state intervention (Harms 2009): leisure activities were used as a tool to reward and stimulate workers, in line with the ideology of the New Man promoted by the Party. Today, leisure serves an urban, "modern and civilized" (*hiện đại – văn minh*) ideal way of life, promoted both by the authorities and land developers, a paradoxical yet typical convergence in post-socialist worlds (Harms 2014). The last type of legacy relates to patterns of thought and social representations: the way wealth is experienced and exhibited today can only be understood in light of socialist legacies. As a social marker of wealth, leisure practices are sometimes minimized in the speech of interviewees who feel suspected of unwarranted practices; but, on the contrary, they can also be put forward as a well-deserved reward. Between visibility and invisibility, these negotiations are characteristic of post-socialist societies, in which individual pleasures expand within the official boundaries tolerated by the party. Therefore, studying the evolution of leisure practices in Ho Chi Minh City may lead to rethinking the legacies of the socialist model in urban Vietnam today (Gorsuch and Koenker 2006; Peyvel 2009, 2015).

In this transitional context, our research question is as follows: how do the different types of legacies from the socialist period play out in the contemporary city as far as leisure practices are concerned? Our main hypothesis is that the development of new urban forms and private leisure places in a globalized context remains articulated with the continuity of urban management forms inherited from the socialist period. Current evolutions lie at the crossroads between the education of the socialist New Man and the globalization of urban models. Indeed, contemporary leisure practices in Ho Chi Minh City increasingly differentiate public and private spaces, collective and individual practices, and public display and intimacy. This hybridity of spaces and practices gives a measure of the reshaped socialist model of leisure since the Đổi mới reforms in 1986.

Our approach focuses specifically on city dwellers' way of life and daily leisure practices, in order to decipher how they are reclaiming the post-socialist city on an everyday basis. Indeed, leisure studies stress how city dwellers have to negotiate their desire for more individual freedom in a post-socialist society, where community used to be a structural value and where the central state was a key actor (Drummond and Thomas 2003). In this transitional period, urban dwellers "domesticate" (Creed 1997) post-socialism and globalization processes through "an everyday and ongoing set of practices that at times are destabilizing but at other times articulate together" (Stenning et al. 2010, 3). Thus, our specific object of study and approach participate in the "ethnographic turn" in global metropolitan studies, which focuses on the strength of everyday urban practices (Roy and Ong 2011).

In this perspective, our conception of leisure is derived from our interview results and identifies the local and vernacular meaning of leisure in the Vietnamese context. Leisure activities are not precisely defined in the Vietnamese language: they are commonly designated through the expressions *đi chơi* (having fun), *giải trí* (clearing the mind), and *thời gian rỗi* (literally, free time). They are rarely associated with dedicated urban places. For instance, badminton is often played in "urban interstices," such as pagoda courtyards, at the border between public and private spheres (Miao 2001). This is a result of the very high population density of this metropolis with 10 million inhabitants, and its lack of open spaces (Gibert 2014). Thus, by "leisure," we consider every activity urban dwellers practice by choice in order to fulfil at least one of the following goals: rest and relaxation, entertainment, and self-development (Dumazedier 1962). Contrary to tourism, these activities take place in everyday space-time (Knafou 1997).

Our study uses qualitative methods. We interviewed thirty young Saigonese people (between twenty and thirty-five years old) and followed them during their leisure practices (yoga, karaoke, and various outings). These people belong to the first generation that grew up within the Đổi mới context and benefited from new political stability and economic growth (Nguyen Quanh and Trinh Duy 2001). Interviews were completed with participatory observation in different places of leisure (bars, golf courses, amusement parks) over extensive periods of time. Our findings address the relation between access to leisure, urban mobility, and the commodification of urban spaces. These are key to understanding the closing, opening, and privatization of places in this post-socialist city.

The structure of our chapter is as follows. First, we focus on the material and governance aspects of the period of transition to post-socialism in Ho Chi Minh City, by enlightening the progressive yet incomplete privatization of places and stakeholders in the leisure sector today. In the next part, we develop the various dimensions of the social and mental legacy of the period of the centrally planned economy with regards to leisure practices. The final part examines political and ideological aspects of the post-socialist legacy by questioning the evolving values associated with leisure practices today. This part demonstrates how leisure participates in reshaping the idea of a model citizen in a globalized time.

From Cultural Houses to "the Crescent": Privatization of Stakeholders, Places, and Leisure Practices

In contemporary Ho Chi Minh City, various stakeholders – both public and private – substantially invest in leisure as an economic sphere. This was not possible before 1986, when the state exercised a strict monopoly on both tourism and leisure organization. State interventionism has been renegotiated with the economic opening of the country, even though the socialist legacy is still visible in many ways today. An examination of the evolution of stakeholders, places, and leisure practices at the time of economic opening reveals the importance of the socialist legacy in understanding the city today.

Understanding the Socialist Legacy

Leisure practices transcend historical periods and political contexts in Vietnam but their cultural and social meanings differ through time and space and often take hybrid forms. For example, during the colonial period, the institutionalization of race courses reactivated the ancient practice of betting on animals (Schweyer 2005; Durand 2011). Once associated with the urban elite, the meaning of leisure practices changed radically under the socialist regime. There was a strong ideological dimension to the project of democratizing a pleasure activity. Leisure was first used to promote the modernist idea of universal progress and the creation of the "New Man." The so-called New Man is defined by three criteria: political (loyalty to the Party), technical (exceptional merit in warfare or productivity), and social (coming from or considered close to the masses). Their typical figures are soldiers, farmers, workers, or even teachers.

After 1975, the leisure sphere in Ho Chi Minh City was developed through two spatial logics. At first, the socialist regime supported the reappropriation and transformation of older colonial places, such as the European Sports Club. During the colonial period this club used to be a place of distinction (Bourdieu 1984; Nguyen-Marshall, Drummond, and Bélanger 2012), accessible to powerful elites through a strict membership policy. After 1975, the new city authorities opened the place up symbolically to Vietnamese workers and employees. They organized matches and events for them, changing both the social meaning and the uses of the place (figure 13.1). The club was then affiliated with 2,000 production units and became very popular with the urban population (Khanh and Tran 1978).

These illustrations show the opening of the once highly selective European Sports Club, inherited from the colonial period. After 1975, the place was opened to workers and employees, who participated in production units' sport tournaments and enjoyed exhibitions, such as the one dedicated to "Hồ Chí Minh City workers' innovations," which attracted more than 200,000 visitors (Khanh and Tran 1978).

Source: Khanh and Tran, 1978.

13.1 The reinvention of the former "European Sports Club" in socialist Ho Chi Minh City.

Second, new places were developed for the specific purpose of leisure, such as *nhà văn hóa* (cultural houses), which were implemented at a very local scale in every neighbourhood and functioned as a kind of community centre. Mass organizations were then in charge of their daily management. At that time, the leisure sphere was not considered a money-making activity, nor even part of the economic sector. Thought of as a tool of social emulation, leisure practices were a reward for the most deserving workers. Leisure was a form of social justice. By guaranteeing democratic access to leisure, the government could also supervise – even control – the urban population during its free time.

The presence of a cultural house in every neighbourhood is part of the socialist urbanism apparatus. These structures are still part of the local cultural and social landscape today. They continue to organize social events for the residents, such as low-cost cooking classes under the supervision of the Women's Union, exhibitions of books or paintings, and political education and information. At the city scale, these local cultural houses are under the direction of "municipal cultural houses" for women or for youth and veterans, and so on. The economic opening of the country after the 1986 reforms did not mean that such collective places disappeared from the city's neighbourhoods: the legacy of the socialist period works together today with processes of privatization and globalization at the city scale.

New Stakeholders under State Control

As many studies of the Đổi mới process show, the 1986 reforms have been associated neither with the "end of the State," nor with an unrestrained and uncontrolled economic liberalization (Gainsborough 2010; London 2014). In the case of the management of the leisure sector in Ho Chi Minh City, we can observe a diversification of both places and stakeholders. This stratification is mainly conducted under the direction of state representatives, as we can see with the influence of the Saigon Tourism organization in the city today. This state organization was created in 1975, and restructured in 1999 under the authority of the Ho Chi Minh City People's Committee to invest in tourism and leisure. The current diversification of its investments – as owner, manager, or investor – illustrates the evolution of the state strategy to secure economic growth in the leisure sector. Saigon Tourism operates in restaurants and hotels, as well as in the fields of transportation, sport, and culture. It remains a leading actor in the sector at the national scale. This state organization is also conducting more and more diversified partnerships with private companies, both Vietnamese and international, to maintain its supremacy.

Saigon Tourism, for example, owns Bình Quới parks – also called "Bình Quới tourist villages" – along the Saigon river in the Bình Thạnh district. The functioning of these parks reveals their ambiguous status, at the interface between public and private sectors. Admission to the parks is free, but many of their services require a fee, such as renting wooden huts or boats on the river, access to fishing activities or to restaurants, and the use of different sports equipment. These parks have become very popular among young couples looking for a green and calm setting to shoot their wedding pictures or teenagers looking for privacy. The parks' setting celebrates a certain nostalgic vision of a rural Vietnam. The Vietnamese Youth Organization often organizes cultural and artistic events in these parks, which illustrates the parks' state management.

In addition to state stakeholders, the contemporary leisure sector in Ho Chi Minh City is structured by "intermediate actors" who operate at the interface between public and private sectors. This is the case of the yoga club called Yoga Living, which has two studios in downtown Ho Chi Minh City. Yoga Living is a private company owned and managed by an entrepreneur in his thirties named Đạt Hoàng, who took advantage of the economic opening. After studying abroad at Wasada University in Japan, Đạt Hoàng decided to give himself the Japanese-sounding pseudonym "Dato" as part of his commercial strategy. He completed his education in management at the Royal Melbourne Institute of Technology (RMIT – the Australian University of Ho Chi Minh City that opened in 2001). He then invested in the real estate sector and opened his own yoga club. Dato has complementary strategies to attract the broadest range of clients. On the one hand, through the club's meticulously designed website and the activation of his personal social network, he aims to attract upper-class clients – both local and expatriates – to whom he offers costly services. On the other hand, he also offers cheaper memberships to Vietnamese working companies, which are offered to their employees through the trade unions inherited from the socialist period.

This example shows the diverse strategies undertaken by private stakeholders to succeed in the post-socialism leisure market, playing both on individual and collective practices. But the simultaneous use of Dato's yoga studios by socially different populations also requires spatio-temporal strategies, with different schedules and classrooms. The public is socially stratified through the different types of membership offered to upper-class individuals and to union members. The same conclusion can be drawn by looking at the structure of the market for cooking classes, quite a popular leisure practice in Ho Chi Minh City today. These classes are offered both by private companies at high prices – many of them dedicated to international tourists and expatriates – and by the Women's Union at much lower prices, also with a different spatial setting and narrative about the activity's meaning. These examples show the importance of the continuity of many socio-spatial structures inherited from the socialist period in the city today, as far as free time is concerned.

Towards a "World Class City": The Growing Privatization of the Leisure Sector

The rise in power of private stakeholders' investments in the leisure and consumption sectors in Ho Chi Minh City goes with a growing privatization of leisure places. This trend is especially visible in the most recent mega-projects, implemented mainly in the outskirts of the city, such as in Saigon South (district 7). This new residential area was constructed at the start of the twenty-first century, under the private management of a Taiwanese-Vietnamese corporation called Phú Mỹ Hưng Development Corporation. The core of this newly urbanized area (*khu đô thị mới*) is occupied both by a massive shopping mall and a riverfront called "The Crescent" after its half-moon shape. Private sport centres also count as central urban amenities of the area. The Crescent area is a semi-public space, open to various publics for free but regulated by specific rules and schedules. It is also characterized by its franchised restaurants and retail chain stores. In such a place, badminton or football matches do not appear randomly on every open space – such as parking lots, pagoda courtyards, or pocket-parks – but take place in a normalized form in dedicated and private places, most of the time invisible to passers-by (figure 13.2). Both the place and the spatial apparatus of leisure practices are renegotiated.

13.2 The privatization and standardization of leisure places in Saigon South.
Source: Marie Gibert.

This planning model illustrates the idea of the "playfulness" associated with what should be a "worlding" metropolis (Roy and Ong 2011) and borrows from globalized and standardized international standards. In this context, the city-state of Singapore, and its Sentosa Island – sold as the "state of fun" by its developers – is a major source of inspiration for Vietnamese authorities. The overseas Vietnamese community (*Việt Kiều*) is also a key operator in supporting the circulation of these new urban models between Vietnam, Asia, and the rest of the world. The production of such places participates fully in the globalization of the Ho Chi Minh City economy today. The Crescent has become one of the emblems of Ho Chi Minh City becoming a "World Class City," with what is seen as high-quality designed leisure places. The multiplication of golf courses in the outskirts of the city also contributes to this image.

Assessing the Post-Socialist Turn: An Approach through Social Practices of Leisure

Negotiating Private Life through Leisure Activities: A Question of Income and Schedule

Post-socialist cities correspond not only to a new economic system, but also to new modes of space production characterized by growing privatization. The development of a salaried middle class constitutes a very important and often underestimated consequence of the emerging post-socialist city (Nguyen-Marshall et al. 2012; Earl 2004). In Ho Chi Minh City today, one out of two inhabitants is a salaried worker. This is a consequence of both the industrialization process and the growth of a service sector. This new body of salaried workers not only has economic and social consequences, but it also has new ways to structure time. Employment contracts – whether on a daily, weekly, or monthly basis – guarantee people regular incomes and allow them to anticipate expenses and to structure a budget, which might include leisure activities. The labour code limits the workday to eight hours and the work week to forty-eight hours. Workers are allowed a weekly day of rest, to which a further dozen days off is added for vacation. This strict division between free time and labour time is the result of a new economic and legal situation.

Regularity is a question of both income and time: it also explains the progressive development of the concept of a weekend among Ho Chi Minh City's urban middle class. According to Patrick Gubry and colleagues' survey of intra-urban mobility in Vietnamese cities, people's leisure practices represent 5.1 per cent of the motivation for mobility for the whole range of social classes (Gubry et al. 2008). This figure should be added to a "visiting family and friends" motivation (8.4 per cent), which leads to visits occurring preferentially during weekends and Wednesdays. This mobility varies according to social class: distraction mobility among rich households is four times more important than for the poorest, and double that of middle-class households (Gubry and Lê Hô Phong 2010). Mobility also varies according to gender: 10.2 per cent of men say they go to a bar every day after work (most of the time at the *bia hơi*, a type of popular café, where beer is cheap), while 0.7 per cent of women adopt this leisure practice.

Free time and private life are strictly linked because leisure is defined not only as playtime, but also as time for individual recreation. Relaxing, playing, or increasing one's cultural knowledge are among the various activities that construct individuality. These activities have already been taken over by an increasing consumer society, which is now tolerated in the post-socialist city context. Thus, practising a leisure activity has a further meaning: it also involves consuming for personal pleasure, for one's own private life. Leisure is now a sphere for personal development, rather than simply a showcase for the socialist ideal of the New Man. Indeed, if people selected by the Party during the period of the centrally planned economy (*bao cấp*) could be rewarded with activities that would qualify today as leisure, it would be an anachronistic misconception to associate them with free time: the selected people did not choose these activities, which were chosen by the state, which controlled every aspects of it (time, place, and type of activity). While practices themselves could be identical with contemporary leisure activities (playing tennis, attending a reading club, etc.), their meaning and the way they are experienced have deeply changed.

Leisure Mobility as Social Mobility

The privatization of leisure places goes hand in hand with their commodification. In other words, money is being charged for access. The result of this economic and social process is spatially manifested in the increasing differentiation of mobility and accessibility. Now just getting to a place is a socially discriminating factor (Truitt 2008). Following interviews with people, we have mapped the spatial perimeters of leisure practices to draw a comparison based on their different places of residence and places of work, and their age, gender, and income. Figure 13.3 focuses on two interviewees: a student, Nhu (traced in black emdashes), and a business man, Nam (traced in black hyphens).

Nhu is a twenty-four-year-old architecture student. In 2010, she came from Can Tho, in the Mekong Delta, to go to university. She lives at her aunt's, in a lower-middle-class neighbourhood. She makes do with the nearby leisure activities, which consist essentially of karaoke singing, shopping with her friends and chatting with them in a café, or sometimes swimming at the public pool. Because she can sometimes use an old motorcycle, she occasionally goes to the city centre, where she drinks iced coffee with friends in fashionable cafes, for example Highland Coffee, a Vietnamese franchise resembling Starbucks. Through this relatively inexpensive leisure practice, she manages to have access to the city centre, where it would be impossible for her to live. She likes to dress up for this occasion. She always chooses busy cafés and enjoys posing for photos taken with her mobile phone. She thus feels like a "real" city dweller, an insider. To show that she has succeeded in this goal, she sometimes posts her photos on Facebook.

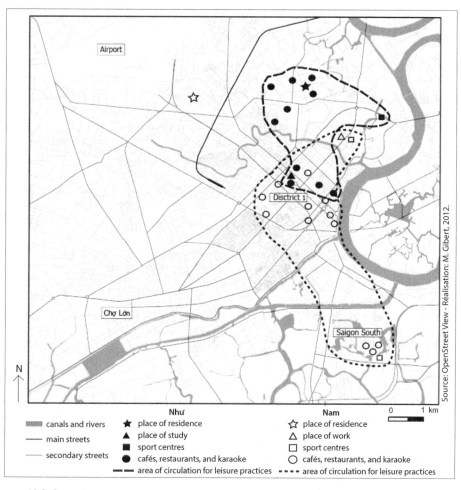

13.3 Spatial perimeters of the leisure practices of two interviewees.
Source: Marie Gibert.

On the other hand, Nam is a thirty-something vice-director of a construction company. He favours the city centre as a place for his leisure activities, especially for going out with his family and friends to restaurants and cafés. But the pollution, the crowds, and the lack of parks and green spaces give him good reason to flee from the city as often as possible. Nam likes going to Saigon South, which he sees as an "urban utopia," built at the start of the twenty-first century ten kilometres away from the city centre, where he enjoys taking a walk with his wife and son. Unlike in the busy heart of Saigon, pushing a stroller here is not a problem. His recent purchase of a car has played a determining role in extending his leisure perimeter. He rarely thinks of using his own neighbourhood as a place to relax in, except for occasional beers at popular cafés with his close neighbours. This is probably a result of the social composition of his neighbourhood, which he chose before moving up in society. Living today in social circles of different levels, through his leisure activities Nam expresses the multiple facets of his upper-middle-class urban identity (Lahire [2001] 2011). He navigates between several social belongings, engaging in high-society practices while embracing other, more popular ones; he articulates his parenthood (while walking with his family) and his masculinity (by frequenting men-only beverage establishments), and his regular attendance at leisure facilities strengthens his knowledge of the city. Those two examples show how leisure mobility and social mobility are deeply linked, and offer insight into the social changes at work in post-socialist cities.

Between Public and Private Spaces: Regulation and Self-Representation of Leisure Activities

The privatization of space goes hand in hand with the privatization of leisure activities and private life. For this reason, the analysis of leisure activities at the domestic scale is enlightening. The first factor involved is obviously the rising standard of living: the more money a household has, the more it can privatize a leisure activity. This was either financially or even socially impossible before, or in certain cases was possible outside the home by sharing the cost of the activity with non-family members. For example, unless a family can purchase a computer, people engage in Internet use and video games at cybercafés. But in other cases, privatization might be the only way to access a leisure activity. Buying illegal DVDs or downloading films allows working-class households to watch movies that would be impossible for them to see in a cinema.

Other leisure activities, such as karaoke, have even more complex spatial trajectories. First invented in cafés, where it was shared between customers, karaoke now involves domestic equipment for the pleasure of a more selected social circle (family, friends, and neighbours). But that does not mean that going out for a karaoke is disappearing. Franchises such as Nice Karaoke now offer private and comfortable rooms, with high-quality equipment, new songs, and amusing competitions. Singing goes hand in hand with other practices of pleasure, such as drinking, eating, joking, and flirting. Young couples, still living in the close quarters of the parental home, enjoy these private rooms for their intimacy, which bears witness to the progressive construction of private life and individual identity.

The complexity of these spatial paths demonstrates that privatization is not the systematic consequence of wealth. Some leisure practices take on meaning only in the public sphere because they require and reveal individual wealth. Ostentatious consumer practices thus bring social rewards because they increase the individual's social capital, and because they allow for a rewarding social life. Leisure practices serve a sort of self-representation. Our interview with Ngan, also known as Alison since she works in TV, brought this process to light. She learned how to play golf primarily for her job and to play with customers. Playing golf can also be seen as a more global strategy of social distinction. She plays in a private club, far from Ho Chi Minh City, which requires a car and an expensive membership, and she likes posting photos and comments on Facebook about her practice. These posts reveal the subtle negotiation between public and private spaces and life in the framework of leisure. They show her complete integration into a place that is normally invisible to nonmembers, and display this status to a social network that is officially blocked in Vietnam. This example shows that leisure is an efficient way to negotiate a gratifying access into spatially and socially secluded contexts, such as golf clubs, tennis clubs, or, to a lesser extent, virtual social networks.

On the one hand, mobility and leisure are a powerful engine for differentiating and hierarchizing social classes, urban and rural areas, men and women, and generations. On the other hand, the labour market links mobility and leisure to values such as employability, adaptability, and open mindedness.

Between Negotiation and Control: Places of Leisure in Post-Socialist Ho Chi Minh City

Becoming a "Civilized Urban Dweller" through Polished Leisure Practices

While leisure can be viewed as an underlying factor for traditional social and bodily norms, such practices also introduce new codes of urban life, in both private and public space. For example, Yoga Living denies late participants access to classes because, as their website once said, "rubber time is not a practice at Yoga Living." In the same way, the banning of eating, chatting, smoking, or phoning in the yoga studio instigates a new understanding of courtesy. Although those rules make leisure practices more rigid, the majority of customers also consider them a quality guarantee. This control of the body, even during "time off," shapes a new urban way of life, conveying a legitimized social and spatial order. Leisure can therefore be viewed as a tool to instil urban, social, and gender codes.

The educational dimension of leisure is particularly visible with children, in and outside of school. In an interview with Tuấn – a forty-five-year-old business man working in the construction sector – we talked about his daughter and son. They both go to the British International School, a prestigious institution, which boasts a large number of activities ranging from sports (volleyball, running, climbing) to theatre, music, photography, and trips (from local hill-stations like Đà Lạt to international destinations like Hong Kong). Tuấn and his wife see these activities as a way for their children to learn how to behave in a civilized manner, and an opportunity for them to be integrated in a changing world where some behaviour and skills that were once considered "useless" are now necessary. He confided to us that he learned swimming very late, with his children, because he thought that not knowing was nowadays not only dangerous but also "ridiculous."

Supervision of young people, even during "time off," was already an important issue before 1986, to form a so-called New Man through mass organizations like the Pioneers. Nowadays, it represents a growing concern for parents belonging to the urban upper and middle class because leisure and education may become more and more complementary in a social- climbing perspective. This explains why leisure for children has gradually become a business, as Hoa, director of the Eveil Centre, explained to us:

> My aim was to open a creative space, where children can express their thoughts. In public schools and in mass organizations for youth, like the Pioneers, teachers impose their rules to children, who are very passive: they copy, repeat, and don't decide. I prefer when animators [the change of term, from teacher to animator, is interesting to note] encourage children to make things by themselves, to participate in the activities, whether it be dancing, drawing, or expressing. The most important is that children acquire technical but also personal skills: not only knowing how to play piano or to dance, but also how to express their opinion, to collaborate and respect their colleagues, to discipline and discover their body. The majority of children have progressed not only in their activities but also at school: they are more active, they have more self-confidence. Parents are very satisfied.[2]

The conjunction between leisure, education, and the labour market is remarkable: creativity, assertiveness, and adaptability are common values, serving the growing individuation of children.

Staging polished and gratifying leisure activities in suitable places is a real issue for the urban middle and upper class. More than the practice itself, it is the place of the practice that matters. That is why it's particularly useful to analyse the post-socialist landscape of leisure – that is, the spatial transformations induced by consumer and leisure society since the Đổi mới policy (Drummond 2012). Carrying out certain practices outside allows people to easily display their wealth. A number of foreign fast foods chains, like KFC, organize birthday parties for children, like Mỹ (a name that means "United States" in Vietnamese), the ten-year-old daughter of one of our interviewees. Such birthday parties contribute to the growing individuation of children. Mỹ's party also allows new distinctive practices, like playing with other children in the playground of the restaurant and eating fried chicken wings and drinking Coca Cola. It also lets her parents recast their success story as directors of a flourishing fertilizer industry in the Mekong Delta. We can develop a similar analysis of gymnastics. The traditional *tập thể dục*, practised very early in the morning on the sidewalk, and the workout practised in the fitness room have the same practice, but in different places that indicate the evolution of the meaning of gymnastics.

Post-Socialist Leisure Practices and the Edification of a "Civilized" City

Local authorities understand leisure as a way to construct an urban ideal. The urban production and success of new private leisure places illustrate the growing convergence of views between urban authorities and the urban middle class. This convergence can be explained both by the will of the authorities to more strictly control recent urban development after decades of laissez-faire, and by the growing concern of the middle class to protect and mark out the boundaries of their newly acquired properties, and also to privatize their leisure places (Harms 2009). There is a growing distinction between public and private urban spaces; once very blurred in the Vietnamese cultural context, the separation between public and private spheres is currently increasing.

Urban authorities themselves officially support this distinction by promoting the intended edification of what is called a "civilized and modern" city (*đô thị văn minh, hiện đại*). Official poster campaigns urge urban dwellers to follow new urban rules of civility, such as not spitting or wearing inappropriate pyjamas (*bộ đồ mặc nhà*) on the street, and not conducting business on the sidewalk – all of which aims to build "cultural neighbourhoods" (*Làm khu phố văn hóa*). If the local inhabitants follow these rules, an official award is displayed at the entrance to their neighbourhood. These urban codes imposed by an authoritarian regime participate in the construction of a dreamed-of new city, at the crossroads between the education of the socialist New Man and the globalization of urban models. The development of new urban forms and private leisure places in a more and more globalized context continues urban management forms inherited from the socialist period.

The similarity between these official boards and the advertising slogans developed by the private managers of new urbanized zones such as Saigon South is also striking. Saigon South promotes, as one of their brochures states, "a community cultural lifestyle in a civilized city." In a post-socialist context, consumer and leisure society participate in a society of political consensus (Drummond and Thomas 2003).

Conclusion

Despite a relatively weak legacy in terms of urban planning and architecture, the case of Ho Chi Minh City illustrates the afterlife of socialist governance. Indeed, the lingering influence of socialist systems continues to affect contemporary "market socialism" and feed increasingly globalized urban life in myriad ways. By navigating everyday spaces and focusing on renewed way of life, studying these interferences through leisure practices allows deciphering the transition to post-socialism. In this context, our study shows the lasting role of the state as a key actor in the sector, despite the progressive diversification of stakeholders and the rise of the private sector. Indeed, the state has adapted its strategy to grasp the economic growth in the leisure sector and keep its control as tight as possible. This highlights Ho Chi Minh City's specific metropolization process in the era of globalization.

Another important aspect of the socialist legacy concerns social representations and practices. Indeed, leisure studies show how city dwellers negotiate in myriad ways their desire for more individual freedom in a post-socialist society (Gorsuch and Koenker 2006 ; Berdahl and Bunzl 2010 ; Banaszkiewicz, Graburn, and Owsianowska 2016), in which the visibility and daily performance of wealth remains ambiguous. Thus, leisure practices involve subtle personal negotiations that oscillate between opening and closing, public and private, street sociability and self-segregation. On a micro level, this approach is an original way to understand the construction of urban post-socialist identities, particularly the way they are renegotiated socially and spatially within a market economy. Through leisure, city dwellers negotiate their right to the city and their access to the city centre, in a more and more globalized context.

Furthermore, leisure remains associated with political ideology: it was at the heart of the edification program of the New Man during the socialist period and continues to participate in the contemporary reshaping of the idea of a model citizen in a globalized time under the slogan of "civilized and modern" citizens. As such, leisure practices accommodate social changes while at the same time reaffirming the role of the central state and its representatives. Thus, the study of leisure offers a renewed perspective on post-socialist Vietnam that should not be reduced to a transitional stage towards neoliberalism.

NOTES

1 *Đổi mới* means "renovation" in Vietnamese and refers to a series of state reforms marking a shift from a centrally planned economy to a "socialist-oriented market economy."
2 Interview, Ho Chi Minh City, 19 April 2013

REFERENCES

Banaszkiewicz, Magdalena, Nelson Graburn, and Sabina Owsianowska. 2017. "Tourism in (Post)socialist Eastern Europe." *Journal of Tourism and Cultural Change* 15, no. 2, 109–21.

Berdahl, Daphne, and Matti Bunzl. 2010. *On the Social Life of Postsocialism: Memory, Consumption, Germany*. Indianapolis: Indiana University Press.

Bourdieu, Pierre. 1984. *Distinction: A Social Critique of the Judgement of Taste*. Cambridge, MA: Harvard University Press.

Creed, Gerald W. 1997. *Domesticating Revolution: From Socialist Reform to Ambivalent Transition in a Bulgarian Village*. University Park: Pennsylvania State University Press.

Drummond, Lisa Welch. 2012. "Middle-Class Landscapes in a Transforming City: Hanoi in the Twenty-First Century." In *The Reinvention of Distinction: Modernity and the Middle-Class in Urban Vietnam*, edited by Van Nguyen-Marshall, Lisa B. Welch Drummond, and Danièle Bélanger, 79–93. Dordrecht, The Netherlands: Springer.

Drummond, Lisa Welch, and Mandy Thomas, eds. 2003. *Consuming Urban Culture in Contemporary Vietnam*. London, UK: Routledge Curzon.

Dumazedier J. 1962. *Vers une civilisation du loisir?* Paris, France: Éditions du Seuil.

Durand, Maurice. 2011. *Imagerie populaire vietnamienne*. Edited by Philippe Papin. Paris, France: EFEO.

Earl, Catherine. 2004. "Leisure and Social Mobility in Ho Chi Minh City." In *Social Inequality in Vietnam and the Challenges to Reform*, edited by Philip Taylor, 351–79. Singapore: ISEAS Publications.

Gainsborough, Martin. 2010. *Vietnam: Rethinking the State*. London, UK: Zed Books.

Gibert, Marie. 2014. "Les ruelles de Hồ Chí Minh Ville (Việt Nam), Trame viaire et recomposition des espaces publics." PhD diss., Université Paris 1 Panthéon – Sorbonne.

Gorsuch, Anne E., and Diane P. Koenker, eds. 2006. *Turizm: The Russian and East European Tourist under Capitalism and Socialism*. Ithaca, NY: Cornell University Press.

Gubry, Patrick, and Linh Lê Hô Phong. 2010. "Niveau de vie et déplacements dans les métropoles vietnamiennes: Hô Chi Minh ville et Hanoi." *Tiers Monde* 201, no. 1: 107–29. https://doi.org/10.3917/rtm.201.0107

Gubry, Patrick, Thi Huong Lê, Thi Thiêng Nguyên, Thuy Huong Pham, Thi Thanh Thuy Trân, and Hoang Ngân Vu, eds. 2008. *Bouger pour vivre mieux: Les mobilités intra-urbaines à Hô Chi Minh Ville et Hanoi*. Hanoi, Vietnam: Editions de l'Université nationale d'économie (NEU).

Harms, Erik. 2009. "Vietnam's Civilizing Process and the Retreat from the Street: A Turtle's Eye View from Ho Chi Minh City." *City and Society* 21, no. 2: 182–206. https://doi.org/10.1111/j.1548-744x.2009.01021.x

– 2014. "Civility's Footprint: Ethnographic Conversations about Urban Civility and Sustainability in Ho Chi Minh City." *Journal of Social Issues in Southeast Asia (SOJOURN)* 2: 223–62.

Khanh, Chi, and Ho Tran. 1978. "Le 'cercle européen' depuis la libération." *Vietnam* 232: n.p.

Knafou R. 1997. *Tourisme et loisirs*. Paris, France: La Documentation française.

Lahire, Bernard. (2001) 2011. *The Plural Actor*. Cambridge, UK: Polity Press.

London, Jonathan, ed. 2014. *Politics in Contemporary Vietnam: Party, State, and Authority Relations*. Basingstoke, UK: Palgrave Macmillan.

Miao, Pu, ed. 2001. *Public Places in Asia Pacific Cities: Current Issues and Strategies*. Dordrecht, The Netherlands: Kluwer Academic Publishers.

Nguyen-Marshall, Van, Lisa Welch Drummond, and Danièle Bélanger, eds. 2012. *The Reinvention of Distinction: Modernity and the Middle Class in Urban Vietnam*. Dordrecht, The Netherlands: Springer.

Nguyen Quanh, Vinh, and Luan Trinh Duy. 2001. *Socio-Economic Impacts of "Doi Moi" on Urban Housing in Vietnam*. Hanoi, Vietnam: Social Sciences Publishing House.

Peyvel, Emmanuelle. 2009. "L'émergence du tourisme domestique au Viet Nam: Lieux, pratiques et imaginaires." PhD diss., Université de Nice-Sophia Antipolis.

– 2015. "Devenir touriste dans un pays socialiste: Le cas du tourisme domestique au Việt Nam." In *La mondialisation du tourisme, les nouvelles frontières d'une pratique*, edited by Emmanuelle Peyvel, Benjamin Taunay, and Isabelle Sacareau. Rennes, France: Presses Universitaires de Rennes.

Roy, Ananya, and Aihwa Ong, eds. 2011. *Worlding Cities: Asian Experiments and the Art of Being Global*. Oxford, UK: Wiley-Blackwell.

Schweyer, Anne-Valérie. 2005. *Le Viêtnam ancien*. Paris, France: Les Belles Lettres.

Stenning, Alison, Adrian Smith, Alena Rochovska, and Dariusz Swiatek. 2010. *Domesticating Neo-Liberalism: Spaces of Economic Practice and Social Reproduction in Post-Socialist Cities*. Chichester, UK: Wiley-Blackwell.

Truitt, Allison. 2008. "On the Back of a Motorbike: Middle-Class Mobility in Ho Chi Minh City, Vietnam." *American Ethnologist* 35, no. 1: 3–19. https://doi.org/10.1111/j.1548-1425.2008.00002.x

14 Mapping Khujand: The Governance of Spatial Representations in Post-Socialist Tajikistan

WLADIMIR SGIBNEV

This chapter delves into the governance of spatial representations of the city of Khujand in northern Tajikistan by means of an analysis of maps and mapping practices – both official, printed ones and hand-drawn mental maps by the city's inhabitants. It retraces the negotiations of spatial representations at the periphery of the former Soviet Union, taking on pre-Soviet, Soviet, and post-Soviet map-making. In doing so, it reveals the divisive, unequal and peripheralizing practices of spatial knowledge production, from the pre-Soviet and Soviet era to independence and after, and thus questions the validity of a qualitative distinction between a socialist and a post-socialist condition within this particular object of analysis. This refers in particular to a continuation of colonial and self-colonializing practices from the pre-Soviet era, through the Soviet period, into the current era after independence. In each period, maps and mapping practices were, above all, employed by authorities as means of domination, and emancipatory mapping practices remain scarce up to today.

The making of socialist and post-socialist subjects through housing or infrastructure (Humphrey 2005; Buchli 2007; Laszczkowski 2016) has received wide academic attention. In the same vein, the study of monumental architecture and ideologies embodied in space is a prominent subject within post-socialist academia (Czepczyński 2008; Forest and Johnson 2011; Koch and Valiyev 2015). This chapter, however, does not attempt to trace the governance of spatial representations through their reflections in the built environment. Instead, it turns to an analysis of the translation process involved in map creation, on the side of production, and map usage, on the side of experience.

Looking at oftentimes conflicting, translation processes, means subscribing to an understanding of space as a social product, one which is being explored and conquered, and seen as "means of control, and hence of domination, of power" (Lefebvre 1997, 26). With regard to Henri Lefebvre's "trialectics" of social space production – that is, the imbrications between conceived, perceived, and lived space (Stanek 2011; for an adaptation of the scheme to central Asian urban contexts, see Sgibnev 2015) – the focus lies here on the *conception* side. Lefebvre has argued that the domain of the conception of space – is primordial: it "dominates over the others, and in its name spatial texture as well as social practice can be changed" (Bertuzzo 2009, 31). This domain is theorized as residing primordially in the hands of experts – town planners, architects and ideology entrepreneurs – who will take up the bulk of the following discussion of the governance of spatial representations. This approach allows me to identify colonializing practices of spatial conceptualization, and their legacies and specificities. The chapter thus contributes to understanding the post-socialist condition from the conceptual side of space production, and their transformations over time; space being a social product, all societal change must be shadowed by a spatial change (Lefebvre 2009, 186), and thus also by changes in the conceptualization of space – including mapping practices.

While maps are valuable tools for orientation in an unknown environment, they also shape our perceptions and conceptions of space, and can therefore have a profound impact on mental maps. If drawing a mental map of Berlin or London, for instance, many might rely on the subway network plan and distort distances, angles, and scales accordingly. Maps lead us to our destinations, but maps can also mislead – not because of a lack of cartographic literacy on our part, but on purpose. In short, maps are situated at the crossroads of space conception and perception – at the crossroads of ideology and everyday routines. Mapmakers want maps to appear to convey "true, probable, progressive, or highly confirmed knowledge," based on "measurement and standardization" (Harley 1989, 4). But focusing on the apparent accuracy of maps we neglect the fact that maps are social constructions, too (Crampton 2001).

Soviet maps in particular triggered the emergence of a critical approach to maps: in the *glasnost* era of the 1980s, the world learned that Soviet topographic maps were falsified on purpose to mislead the enemy (Unverhau 2006; Postnikov 2002). The ensuing public outcry was based on the fierce conviction that "Western maps are value free," objective, and truthful (Harley 1989, 5). Questioning this stance led to the emergence of a "critical cartography" (Glasze 2009) that transcended the front lines of the Cold War. Maps thus appear as (re)productions of social orders and instruments of domination, as I will point out throughout the following sections.

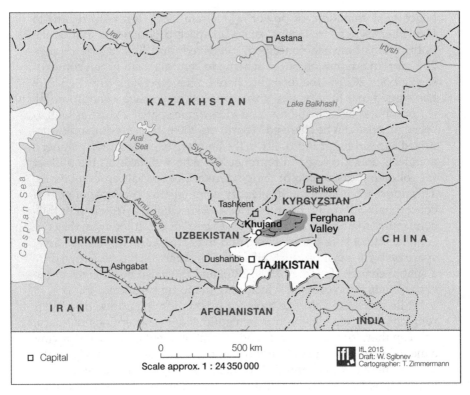

14.1 Map of Tajikstan. IfL 2015. Draft: W. Sgibnev. Cartographer: T. Zimmermann.

The Mapped Object: Khujand and Its Built Environment

I will be looking at the city of Khujand, where the fieldwork for the present chapter took place in 2009, 2010, and 2014. The city is located in the Tajik part of the Ferghana valley in Tajikistan's northernmost part. The ethnically diverse city has roughly 160,000 inhabitants, out of an agglomeration of some 750,000 inhabitants, which makes it the second-largest in the country. It became part of the Russian Empire in 1866 and, thanks to its early industrialization, was the cradle of virtually all leaders of the Tajik Republic in Soviet times (1917–91).

The wide Syrdarya River forms the dominant east-west axis of Khujand's layout.[1] One large street – Lenin Street – runs from the southeast all the way from what was the largest silk plant in the Soviet Union's right to the foot of the Mogol-Tau Mountains. The city's main institutions are located on both sides of Lenin Street – the sprawling Panjshanbe bazaar and the city's main mosque with the shrine of Shaykh

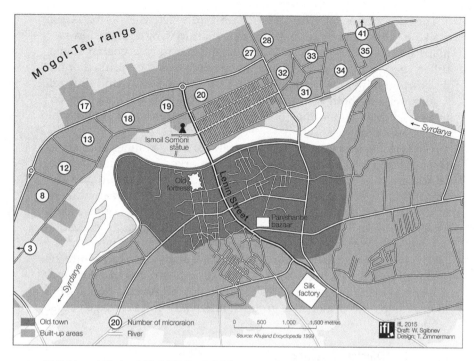

14.2 Map of Khujand. IfL 2015. Draft: W. Sgibnev. Cartographer: T. Zimmermann.

Maslihiddin, and the central stadium and the administrative buildings closer to the bridge. Not far off Lenin Street, one finds the theatre, the post office, and the reconstructed old fortress with the regional museum. Right behind the fortress, the central park spreads out northward towards the river.

On the southern shore of the Syrdarya River – that is, on the left bank – we find densely built-up quarters with an irregular street layout. This is Khujand's old town. After crossing the Syrdarya Bridge, Lenin Street runs uphill to the university campus, past the Ismoil Somoni statue that came to replace the Lenin statue in 2011. The city used to bear Lenin's name from 1936 to 1992, when it was called Leninabad. On the right bank, the city looks very different. There are no more detached houses, no more crooked streets. Here, the Soviet Union has left behind its most lasting built heritage – the *microraions* or microdistricts, large housing estates consisting of prefabricated multi-storey houses. On the fringes of the microraions, rusting industrial estates recall another Soviet heritage: the large-scale industrialization of central Asia followed by the collapse of industry after the breakdown of the Soviet Union (see Sgibnev 2011).

This description of Khujand from a bird's eye view is based on walking – on spatial experience rather than looking at maps. In fact, maps are rather hard to come by in Khujand and people's perception of urban space comes without being too much troubled by printed or on-line cartographies. Nevertheless, maps of Khujand exist, and they tell a story relating both to underlying conceptualizations from "above" or from abroad, and to the "unmapped" spatial representations of the city's population.

The system of orientation relies heavily on daily routines and modes of mobility and is therefore intrinsically marked by gender, income, education, and cultural and linguistic background (Istomin and Dywer 2009, 40–1). Systems of orientation link up with the mental maps of Khujand's inhabitants, which I analyse. The information is drawn from a series of interviews, participant observation, and a collection of some eighty mental maps, complemented by archive material.[2] My interview partners drew their own mental maps on plain white A4 sheets of paper, mostly towards the end of semi-structured interviews relating to everyday experiences of urban space.[3] An additional set of mental maps was drawn by a group of university and language-course students.

Khujand in "Official" Cartography

Khujand looks back on a long-standing history of cartographic display. Since the city is, presumably, the location of the ancient Alexandria Eschate – the "farthest Alexandria" reached by the Macedonian conqueror – both antique and medieval maps mention it as well as the surrounding Sogdiana region. However, these historical maps involved a rather exotic projection of European ideas right into the heart of Asia, and none of them made it to the city itself. As for local cartographic production, we have to mention Arab-language material, both from the central Asian oasis as well as from travelling merchants and scientists, such as Ibn Hawqal and Al-Istakhri (both from the second half of the tenth century). Postnikov (2007, 48) argues that their maps were "strikingly different from European" ones, being "graphical descriptions of … elements, which were subject to ideas of symmetry and geometrical conventions prevalent in the Arab Caliphate's cosmography and visual arts." Nonetheless, we can barely consider these as an expression of "indigenous" spatial knowledge.

The next step in mapping spatial knowledge of Khujand and its surroundings came with the expansion of the Russian empire. From the seventeenth to the nineteenth century, cartographic production both laid the ground for the incorporation of central Asian territories and solidified the empire's hold on them (Postnikov 2007, 55). At the same time, British "counter-mapping" endeavours played their role in the "Great Game" between both empires for the central Asian "heartland," which they considered a cornerstone for the domination of Asia.

Finally, a proliferation of maps on a larger scale occurred with the rise of Soviet power and the ensuing industrialization and rise in literacy. The former necessitated detailed cartographic surveys for resource extraction and territorial planning; the latter expanded the number of specialists available to carry out these tasks, and a wider ability to "read" maps in the general population. However, map usage largely remained restricted to specialists, and the use of maps on an everyday basis remained scarce. In this aspect, Soviet cartographic practice inherited and exacerbated the Russian Empire's "centralized management of cartography" (Postnikov 1996, 168), featuring "draconian restrictions imposed on compilation, publication and use of large-scale maps" (169).

Furthermore, if maps were at all available, their content was often doubtful. Borén and Gentile (2007, 101) attribute this to "the real socialist fear of accurate information." Detailed topographical maps were in most cases intended for military purposes (and geological surveys, to some extent), and therefore were classified and unavailable to the general public. The case was even more difficult with detailed city maps, which were not available at all. Instead, tourist maps were sold, though they were conceived more as souvenirs than as an aid to orientation. In fact, individual tourists were rare, and participants in an organized group tour with a guide did not necessarily need maps. Soviet tourist maps were therefore not to scale, "starkly reduced and had limited circulation; they were disproportionate and incomplete" (Pilz 2011, 82). Smaller residential streets did not appear on these maps. However, far-away suburbs could be included, as long as they had representative value, as Pilz (2011, 86–8) observed of maps of Tbilisi.

The Khujand/Leninabad area was, like other parts of the Soviet Union, covered by detailed topographical maps, down to a scale of 1:500,000, but these were classified and therefore beyond public reach.[4] Because Leninabad was a closed town for most of its existence, tourist groups were not as common as for instance in the Republic's capital Dushanbe.[5] This also meant that people were not allowed to take photographs from above the second storey of a building, which explains why very few panoramic photographs of Leninabad exist, not to speak of aerial images.[6]

The only city map of Leninabad that I found in the public library or in private collections is a 1982 tourist map.[7] It shows the major streets and quarters on both sides of the Syrdarya River, as well as the main monuments, sights, museums, libraries, and theatres. The available information on the city centre is summarized, and the street alignment is rather sloppy, but it is still useful for locating the major streets of the old town, for instance. However, the map's most astonishing feature is that it is wrong about the mass housing estates on the right bank. Instead of being aligned in a row running east-west, they are shown as arranged in two diagonally arranged blocks. Consequently, all the streets, landmarks, and blocks on the right bank are terribly misrepresented. I cannot explain this distortion, since Marafiev, who served as consultant for that map, is a historian and expert of Khujand's quarters. Although this map was inaccurate it did not do much harm since its circulation was very limited. But it serves as a good example of the Soviet policy of knowledge propagation, and, generally speaking, the ways that maps "extend and reinforce the legal status, territorial imperatives and values stemming from the exercise of political power" (Harley 1989, 12).

The Soviet cartographic policy remained largely unaltered after Tajikistan's independence in 1991. The most detailed map of Khujand available today is an annex to the *Khujand Encyclopedia* (Abdullaev 1999), drawn by hand by Khudoyberdiev, probably tracing a declassified topographical map.[8] He has taken some liberties as far as the industrial areas on the right bank are concerned, but still, this map remains the most reliable and accurate printed map of the town. It is relatively expensive because it is only available as part of the encyclopedia, but it is still for sale in both of Khujand's major book stores. The map measures 20 × 30 centimetres in its paper version, though I have also seen it as a digitally enlarged A0 print[9] in one real estate agency and one government institution – which once again underlines that this is the most accurate map on the market. The city planning department boasts on its wall a detailed Soviet-era map of the city, which also serves as bottom layer for a series of planning documents. Despite no longer being up-to-date, the map is still considered classified and cannot be photographed or digitally reproduced.

Bill (2010, 243) reports that a new map of Khujand is on the drawing board, but is not yet available. Other maps exist, but are never seen in public. The 2000 Lonely Planet guide to central Asia boasts a map of Khujand (Mayhew, Plunkett, and Richmond 2000, 443), which is even significantly less detailed than the Soviet 1982 map. It does not show much more than Lenin Street and some buildings alongside of it, and completely omits the rest. A detailed map of Khujand is found on the rear of the 2008 Orell Füssli map of Northern Tajikistan (Northern Tajikistan 2008), yet it is rather awkwardly framed so that it leaves out a large western chunk of the city – that is, the entire old city – and omits to number the mass housing estates. This renders the map much less useful than it could be. The map also seemingly served as the basis for the one provided in Bill's recent guidebook on Tajikistan (Bill 2010, 198). The 2009 tourist map of Tajikistan by Map Factory features a detailed and colourful map of Khujand – the best map to my eyes, had they numbered the microraions on the right bank.[10] Yet it was on sale only in Dushanbe at the time of my field research and was therefore unknown in Khujand.

Other, more widely available, cartographic products do not feature city maps at all – such as the 2004 school atlas – and therefore have no detailed reference to the Khujand urban area. The school atlas mostly consists of reprint maps of the Soviet 1968 atlas of Tajikistan, and the 1974 encyclopedia of the Soviet Tajikistan, bearing witness to the very scarce cartographic production since independence. Another cartographic product frequently encountered in NGOs and state administrations is the World Bank-sponsored *Socio-Economic Atlas of Tajikistan*, freely available online.[11] Again it has no details of Khujand. What strikes the reader is the amalgamation of two areas of arcane state knowledge: official statistics generally are as scarcely available and as unreliable as maps.

Also striking about the official usage of cartographic material is its incorporation into monuments. In Khujand, there are three major examples of this phenomenon. One is painted on a large sign located close to the entrance of the city park in front of the reconstructed walls of the old fortress and shows the Syrdarya River and the extent of Khujand's medieval fortifications. If there is one map that has had an influence on the mental maps of Khujand's inhabitants, it is this one, which reminds the citizens of the city's long-standing and mythified history. The second example is a map set in red marble at the monument for Kamoli Khujandi, a fourteenth-century poet and traveller. It boasts his itineraries from Khujand through Tabriz to Mecca, thus supporting the claim of Tajik/Persian being the primordial language of culture and literature for the entire region. The Ismoili Somoni monument that came to replace the Lenin statue in 2011 boasts a colourful mosaic depiction of the Somonid Kindgom of the eleventh century, which at that time encompassed most of current central Asia and is now officially proclaimed as the precursor of Tajik statehood. This third example of maps on monuments has its prominent counterpart on an earlier map of the Somonid Kingdom at the Ismoili Somoni monument in the capital Dushanbe, erected in 1999. The current borders of Tajikistan serve to compare past grandeur with contemporary marginalization, and support irredentist claims on the "Tajik" metropolises of Samarkand and Bukhara in contemporary Uzbekistan.

It seems that keeping maps rare for the general public strengthens their credibility when being displayed in official contexts – or at least the authorities' conviction that this should be the case.[12] Herfried Münkler's (1995) theory of the "reserve of visibility" seems to be at work here. The practice of map production and map use is very much top-down rather than collaborative, and is in many ways reminiscent of Soviet modes of managing spatial knowledge. The maps I discuss reflect what Dodge and Perkins (2015, 38) have called a "conspiratorial view of how space is governed for the benefit of the few." To counterbalance this very state-centred view of spatial knowledge production, I turn to yet another set of Khujand's maps – this time not "official" ones, compiled and printed by various authorities, but the mental maps of Khujand's inhabitants. They illustrate some of Khujand's most important features and explicit local patterns of space conception and perception.

Khujand's Mental Maps

Low (1996, 384) argued that "the city as a site of everyday practice provides valuable insights into the linkage of macroprocesses with the texture and fabric of human experience." Mental maps have proved valuable tools to discern these everyday practices and link them up to a politicized and problematized analysis of spatial conceptualizations. Downs and Stea (1982, 107) insist on the fact that mental maps emerge from social interaction, from a process "that can cope with both social requirements and individual needs." This interaction is indeed subject to power relations, thus turning mental maps into a tool for discerning modes of governance of spatial representations. Lefebvre, who was preoccupied with a critique of everyday life, would probably support the proposal for a similarly critical approach to representations of everyday life – that is, a reflexive and sensitive interpretation of mental maps that takes into account underlying power relations. In the same way as interviews, mental maps are culturally sensitive – both in regard to the local knowledge of the interview partners who sketch them, and to the researcher in his or her efforts of interpreting them (Feinberg et al. 2003, 251; Majewski, 2010).

Compilation and exchange of spatial data allows people to cope with the complexity of the world that surrounds them and produces inter-subjectivity with regard to the production of space. In the convergence between geography and psychology, mental maps gained prominence as tool for grasping patterns of space perception and thus contribute to citizen-oriented spatial planning, first of all thanks to the works of Kevin Lynch (1960). Building on the contribution of Downs and Stea (1982), mental maps also increasingly became a tool for gaining insight into underlying spatial conceptualizations and processes of appropriation of space. Spatial knowledge thus appears not just as passively learned: through co-producing spatial knowledge, individuals are actively producing space (Downs and Stea 1982, 23; Ploch 1995, 26). Mental mapping is an intersubjective and dynamic process, embedded in social structures and practices. In fact, spatial knowledge is a central component of individual identity (Weichhart 1990). Looking into this "sense of place" (Downs and Stea 1982, 49) reveals insights into practices and processes of spatial knowledge production, as well as underlying patterns of power and domination. In this regard, mental maps meet the call of critical geographers to "look beyond the power of material artefacts ... onto *mapping* and the numerous practices that bring mapping into being ... toward the study of personal map spaces and their co-construction, focusing on everyday practices" (Dodge and Perkins 2015, 38, emphasis in the original).

Contrary to "official" maps that focus on the representative dimension or conceal classified information, everyday life forms the basis of people's individual mental maps. For example, in people's sketches, the Syrdarya River appears as one major landmark, which is a unique feature of Khujand compared to other central Asian cities – this is a feature that mental maps share with their "official" counterparts. Almost all respondents also mentioned the central Panjshanbe bazaar. The importance of the bazaar is consistent with the ideal-type topology of a central Asian city (Wirth 2002, 151).

However, it is striking that respondents constantly omit the old city with its maze of crooked streets, mosques, and ponds. Even people who live in the old town do not draw it and prefer to stick to the central street and a couple of landmarks. It seems a dramatic discrepancy between the old town's structural and social significance, and its omission in mental maps. This omission might be explained by the largely negative qualities that are commonly ascribed to the old city. The domination of negative terms began with the Russian conquest of central Asia and was exacerbated under Soviet rule (Stronski 2010, 22ff.). The image of danger and filthiness that the Russian colonial administration imprinted on the "indigenous settlements" stuck and grew in Soviet times. After independence, a re-valourization of the old city did not take place, in spite of a general recourse to history for nation-building purposes. A presumably Western urban modernity, characterized by high-rises along broad streets, remains the dominant paradigm. Pres-ervation of historic sites was uniquely aimed at landmarks, such as the Shaykh Maslihiddin shrine. As Pincent (2009, par. 28) argues, "the ur-ban (built) heritage is a state concept which is used for economic and political needs, not for historic ones. It aims above all at monuments which are useful to political authorities." It appears that inhabitants do not consider the old town quarters to be truly "urban" – not "repre-sentative" enough to be talked about as part of the "city." This omission is not only an effect of mental maps, which tend to omit negatively connoted spaces (Ploch 1995, 28), but seemingly runs deeper into their conception of urban space.

With printed maps only scarcely available, the value of personal, gendered, socially and culturally stratified experience for orientation is even higher than general. Ploch (1995), for instance, argues that women tend to sketch areas closer to home and include more details such as shops and social infrastructure. Men tend to sketch their workplace. Furthermore, women often omit unbuilt space, from which Ploch con-cludes that these spaces are linked with insecurity and thus with fear. In general, men's mental maps are characterized by a wider grasp, an attitude of self-evident mastership of space (Ploch 1995, 32–3). Khu-jand's mental maps are no exceptions. On one young woman's men-tal map, we see only the most representative buildings along the main street – the Second World War monument, the bazaar, and the univer-sity. The workplace – and even more so the home – of a young woman must not be revealed. The mental map of an older woman, however, is very much home-centred. It features only the immediate surroundings of her apartment building and the adjacent Yagodka quarter, which she says she likes a lot.

Borén and Gentile (2007, 101) argue that "given that most urban com-munal and commercial services such as shops [and factories and other objects] tended to stay under conditions of central planning, the inhab-itants of socialist *urbs* generally knew their city better than their capi-talist city counterparts." Pilz (2011, 81) endorses this point of view and proposes that in the post-Soviet city orientation "is made on the basis of general directions and conspicuous landmarks of the city such as shops, house colors, oriels, towers, monuments, etc. ... this system plays with fine nuances and presupposes a lot of local knowledge."

This might explain the use of imaginative and catchy toponyms one might encounter in Khujand (and elsewhere in central Asia). Local place names are also fostered by the rise of the *marshrutkas* – the ubiqui-tous minibuses that carry the majority of Khujand's passengers – which are not required to stop at predefined stops. The passengers need to clearly designate intermediate locations between road intersections. In Khujand's eighteenth microraion, marshrutkas generally stop three times along one 400-metre-long snake-like building; first at its beginning, coming from the centre, at "Sari dom" – "start of the house"; then "Tonnel" – a passageway underneath; and finally "Asal" – "Honey," for the name of a shop in front of it. This system of orientation might also explain why new street names do not catch on: it is less a question of ideology than nostalgia; the old names are just too necessary for orientation without maps.[13]

Digital Mapping Practices

Dodge and Perkins (2015, 38) have criticized Harley (1989) for his neglect of the "possibilities of counter-mapping and more inspiring examples that suggest ways of resisting hegemony." Such counter-mapping possibilities are still not taking shape in Khujand. The digital divide plays a role: infrastructure penetration is still poor, though steadily growing (Atoev 2009). Within the country, online access is very unequally distributed, given dire electricity provision regimes in the countryside and often problematic network coverage outside of agglomerations (OpenNet Initiative 2010).

The digital divide makes itself very harshly felt not only regarding access to online services, but also with regard to their content. Tajik-language content is relatively scarce, and most services are available in Russian language only (Atoev 2009). While Russian language proficiency is decreasing among the young population (OpenNet Initiative 2010), it still dominates in the region's online content. When it comes to cartographic resources, Yandex Maps – the most popular resource for the Russian-speaking online world – does not cover Tajikistan in detail.[14] Google Maps has an astonishingly detailed coverage of Khujand, and is the only resource providing details of the old city.[15] However, it is not a popular resource in the country and therefore goes widely unnoticed. Potentially, the growing use of smartphones and the spread of the Android platform which has Google's map service incorporated into it might change the situation. The collaborative and open-source OpenStreetMap coverage is still very rudimentary.[16]

Looking at collaborative map mashups – that is, a combination of proprietary or open-source map layers with additional, crowd-sourced information – one sees they are equally ambiguous.[17] On the one hand, we find the collaborative platform Wikimapia, which shows some high mapping activity for Khujand with several hundreds of objects named on a detailed satellite bottom layer.[18] However, even this bottom-up

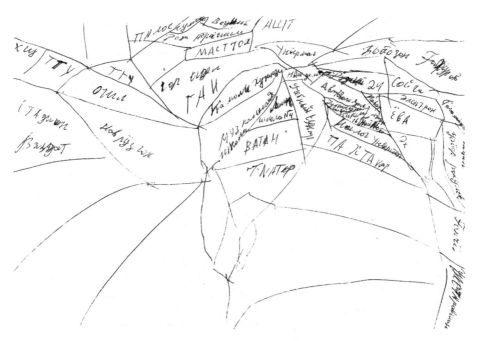

14.3 Map hand-drawn by an interviewee in Khujand, 2009.

cartography seems to reproduce the neglect of the old town. In Wikimapia, this might be because most mapping activity in Khujand is performed by Russian-speaking emigrants who left Khujand after the collapse of the Soviet Union. With equipment and knowledge on "this side" of the digital divide, they exhaustively map places they fondly remember – factories, housing estates, schools, and school excursion destinations. However, they omit the old town, mosques, and tea-houses that were not part of their living world, although they are central to those currently living in Khujand. As for the old town, there might be practical reasons to omit its details: the small and winding lanes are difficult to depict and data are very scarce. In my view, this once again illustrates the low status of the old town – matching the way it appears (or rather does not appear) in the mental maps collected in Khujand. In this vein, the Soviet conception of space gains momentum in contemporary mapping. The printed maps have probably not influenced this, because of their limited circulation, but they parallel this state of mind on another level.

On the other hand, "local" mashup production remains rudimentary. As of 2014, only one mashup application existed that specifically targeted the city of Khujand. A local blogger set up a "city problems" platform using a Yandex map application programming interface.[19] His idea was to provide an interface between citizens and city administration to report various sorts of problems (potholes, heaps of rubbish, leaking pipes) and have them quickly fixed to "make the city better."[20] The announcement circulated widely, and to large acclaim, but only a handful of "problems" have been reported. A wider spread of online cartography is within reach thanks to growing digitalization. However, the limited cartographic literacy of Soviet and post-Soviet times might be an obstacle to rapid development. Digital cartography's promise of "socially progressive" mapping to overcome the "era of colonial mapping" (Dodge and Perkins 2015) seems to be as far as ever on the post-Soviet periphery.

Conclusion

In this chapter I have argued for a politicized view of cartographic productions, looking at maps and mapping practices as "as powerful indicators of nationalism, self-image, attitudes and aspirations" (Downs and Stea 1982, 83). As official representations of space, maps reflect the dominant conceptions of space. Their absence in Khujand contributes to the importance of an immediate experience of space for orientation, and therefore provides a particular structuring of space perception. The omission of the old city can be retraced on official maps, mental maps, and in digital mapping – as well as in official tourist guides – that could have been a valuable resource on the old city. Thus, the Soviet and contemporary official penchant for monumentalism, and the disregard of everyday spatial practices, is felt in all channels of cartographic production, including mental maps. This "colonial" yet deeply internalized conceptualization of space is reproduced through emigrants' mapping.

In the light of maps and mapping, Khujand's post-socialist conditions come to the fore as a melange of Soviet cultural techniques, colonial and self-colonializing practices, and ongoing inequalities after independence – both on a global scale, when we think of the digital divide, yet also on a local level when talking about individual language or technical skills which may be, in Khujand as elsewhere, subject to class divides. Therefore, even if we might find a particular post-socialist melange of practices bound in space and time, the governance of spatial representations has remained highly skewed towards particular political elites with particular modernization agendas. Consequently, this defies a clear-cut distinction between a socialist and a post-socialist mode of the governance of spatial representations. This brings us back to Lefebvre (2009, 206) who argued that "state socialist" and "state capitalist" modes of space production do not differ, yet both rely on exploitative centre-periphery relations, the reproduction of which we witness in independent Tajikistan.

NOTES

1   This makes the place rather special because there are few central Asian cities immediately situated on the banks of a major river.

2   The field research in Khujand, on which the present chapter is based, was generously supported by the German National Academic Foundation. This chapter partly relies on literature and data described in more detail in my PhD thesis (forthcoming in the "Beiträge zur Regionalen Geographie" series).

3   Approximately letter-size in North America.

4   *Northern Tajikistan 1:500,000, Tourist Map of Sughd: With Adjacent Areas of Uzbekistan and Kyrgyzstan* (Zürich, Switzerland: Orell Füssli Kartographie AG; Gecko Maps [distributor], 2008).

5   Interview with a photographer, Khujand, September 2010.

6   Interview with a photographer, Khujand, September 2010.

7   Ûnusova, L.M., S.Š. Marafiev, and L.G. Pilipenko. *Leninabad* [map] (Glavnoe Upravlenie geodezii i kartografii pri Sovete Ministrov SSSR, 1982).

8   In more general terms, compiling encyclopedias was a major task for administration and academia throughout central Asia in the 1990s, both to contribute to nation-building endeavours and to affirm local identities.

9   ANSI E, or 34 × 44 inches.

10   *Tajikistan tourist map 1:200,000* (Dushanbe, Tajikistan: Map Factory, 2009).

11   University of Southampton in collaboration with Tajikistan State Statistical Committee, *Socio-Economic Atlas of Tajikistan 2005*, The World Bank, https://siteresources.worldbank.org/INTTAJIKISTAN/Resources/atlas_11.pdf

12   Other examples working in this direction are, for instance, globes erected in Dushanbe and Tashkent where only one country – Tajikistan and Uzbekistan, respectively – are shown, covering a major part of the earth's surface, and are meant to exalt the grandeur of the country on the global scale.

13   Publishing a detailed city atlas with new names might help the authorities to entrench the new names in popular conscience. Such an atlas would provide the new names with an orientational authority and decrease the need for the "traditional" system of orientation. This happened in Bishkek and Osh, in Kyrgyzstan, and it seems to work.

14   Map of Tajikistan by Yandex.ru, 2015, https://maps.yandex.ru/10322/khujand/?ll=69.605580%2C40.280775&z=14

15 Map of Khujand by Google Maps, 2018, https://www.google.de/maps
    /@40.2817241,69.6255979,13.75z
16 Map of Khujand by OpenStreetMap, 2015, http://www.openstreetmap
    .org/#map=14/40.2871/69.6177
17 For a detailed analysis of the development of mashup services using
    Google Maps, see Dalton 2013.
18 Map of Khujand by Wikimapia, 2015, http://wikimapia.org/#lang=de&
    lat=40.282800&lon=69.614096&z=14&m=b
19 http://khujand-city.tj/problems, accessed 23 March 2018.
20 Parviz, "Kak sdelat' gorod lučše?" [blog post], 27 Juy 2014, http://
    blogiston.tj/root/kak-sdelat-gorod-luchshe/

REFERENCES

Abdullaev S.A., ed. 1999. *Hudžand: Ènciklopediâ.* Dushanbe, Tajikistan: Glavnaâ naučnaâ redakciâ tadžikskih ènciklopedij.

Atoev, Asomudin. 2009. *Access to Online Information and Knowledge: Tajikistan Country Report.* Johannesburg, South Africa: Association for Progressive Communication. http://www.giswatch.org/country-report /tajikistan#attachments

Bertuzzo, Elisa T. (2009). *Fragmented Dhaka: Analysing Everyday Life with Henri Lefebvre's Theory of Production of Space.* Stuttgart, Germany: Steiner.

Bill, Sonja. 2010. *Tadschikistan: Zwischen Dušanbe und dem Dach der Welt.* Berlin, Germany: Trescher.

Borén, Thomas, and Michael Gentile. 2007. "Metropolitan Processes in Post-Communist States: An Introduction." *Geografiska Annaler, Series B: Human Geography* 89, no. 2: 95–110. https://doi.org/10.1111/j.1468-0467.2007 .00242.x

Buchli, Victor. 2007. "Astana: Materiality and the City." In *Urban Life in Post-Soviet Asia,* edited by Catherine Alexander, Victor Buchli, and Caroline Humphrey, 40–69. London, UK: University College London Press.

Crampton, Jeremy W. 2001. "Maps as Social Constructions: Power, Communication and Visualization." *Progress in Human Geography* 25, no. 2: 235–52. https://doi.org/10.1191/030913201678580494

Czepczyński, Mariusz. 2008. *Cultural Landscapes of Post-Socialist Cities.* London: Ashgate.

Dalton, Craig M. 2013. "Sovereigns, Spooks, and Hackers: An Early History of Google Geo Services and Map Mashups." *Cartographica: The International Journal for Geographic Information and Geovisualization* 48, no. 4: 261–74. https://doi.org/10.3138/carto.48.4.1621

Dodge, Martin, and Chris Perkins. 2015. "Reflecting on J.B. Harley's Influence and What He Missed in 'Deconstructing the Map.'" *Cartographica: The*

International Journal for Geographic Information and Geovisualization 50, no. 1: 37–40. https://doi.org/10.3138/carto.50.1.07

Downs, Roger M., and David Stea. 1977. *Maps in Minds: Reflections on Cognitive Mapping.* New York, NY: Harper and Row.

Feinberg, Richard, Ute J. Dymon, Pu Paiaki, Pu Rangituteki, Pu Nukuriaki, and Matthew Rollins. 2003. "'Drawing the Coral Heads': Mental Mapping and Its Physical Representation in a Polynesian Community." *The Cartographic Journal* 40, no. 3: 243–53. https://doi.org/10.1179/000870403225012943

Forest, Benjamin, and Juliet Johnson. 2011. "Monumental Politics: Regime Type and Public Memory in Post-Communist States." *Post-Soviet Affairs* 27, no. 3: 269–88. https://doi.org/10.2747/1060-586x.27.3.269

Glasze, Georg. 2009. "Kritische Kartographie." *Geographische Zeitschrift* 97: 181–91.

Harley, Brian J. 1989. "Deconstructing the Map." *Cartographica* 26, no. 2: 1–20.

Humphrey, Caroline. 2005. "Ideology in Infrastructure: Architecture and Soviet Imagination." *The Journal of the Royal Anthropological Institute* 11, no. 1: 39–58. https://doi.org/10.1111/j.1467-9655.2005.00225.x

Istomin, Kirill V., and Mark J. Dwyer. 2009. "Finding the Way: A Critical Discussion of Anthropological Theories of Human Spatial Orientation with Reference to Reindeer Herders of Northeastern Europe and Western Siberia." *Current Anthropology* 50, no. 1: 29–49. https://doi.org/10.1086/595624

Koch, Natalie, and Anar Valiyev. 2015. "Urban Boosterism in Closed Contexts: Spectacular Urbanization and Second-Tier Mega-Events in Three Caspian Capitals." *Eurasian Geography and Economics* 56, no. 5: 575–98. https://doi.org/10.1080/15387216.2016.1146621

Laszczkowski, Mateusz. 2016. *"City of the Future": Built Space, Modernity and Social Change in Astana.* New York, NY: Berghahn.

Lefebvre, Henri. 1997. *The Production of Space.* Translated by Donald Nicholson-Smith. Oxford, UK: Blackwell.

– 2009. "Social Product and Use Value." In *State, Space, World: Selected Essays,* edited by Neil Brenner and Stuart Elden, 185–95. Minneapolis: University of Minnesota Press.

Low, Setha M. 1996. The Anthropology of Cities: Imagining and Theorizing the City. *Annual Review of Anthropology* 25: 383–409.

Lynch, Kevin. 1960. *The Image of the City.* Cambridge, MA: MIT Press and Joint Center for Urban Studies.

Majewski, Günter E.H. 2010. *Die Weltkarte als "mental map": Die Überwindung des eurozentrischen Mercator-Weltbilds in der Kartographie und im Erdkundeunterricht durch die Peters-Projektion.* Bruchsal, Germany: Verlag für Kulturwissenschaften.

Mayhew, Bradley, Richard Plunkett, and Simon Richmond. 2000. *Central Asia.* 2nd ed. Hawthorn, Australia: Lonely Planet.

Münkler, Herfried. 1995. "Die Visibilität der Macht und die Strategien der Machtvisualisierung." In *Macht der Öffentlichkeit-Öffentlichkeit der Macht*, edited by Gerhard Göhler, 213–30. Baden-Baden, Germany: Nomos.

OpenNet Initiative. 2010. *Tajikistan Country Report*. https://opennet.net/research/profiles/tajikistan

Pilz, Madlen. 2011. "Tbilisi in City Maps: Symbolic Construction of an Urban Landscape." In *Urban Spaces after Socialism: Ethnographies of Public Places in Eurasian Cities*, edited by Tsypylma Darieva, Wolfgang Kaschuba, and Melanie Krebs, 81–105. Frankfurt am Main, Germany: Campus.

Pincent, Guillemette. 2009. "Le patrimoine urbain en Asie centrale." *EchoGéo* 9 [online]. http://echogeo.revues.org/11220

Ploch, Beatrice. 1995. "Eignen sich Mental Maps zur Erforschung des Stadtraumes? Möglichkeiten der Methode." *KEA – Zeitschrift für Kulturwissenschaften* 8: 23–41.

Postnikov, Alexei. 1996. *Russia in Maps: A History of the Geographical Study and Cartography of the Country*. Moscow, Russia: Nash-Dom – L'age de l'homme.

– 2002. "Maps for Ordinary Consumers versus Maps for the Military: Double Standards of Map Accuracy in Soviet Cartography, 1917–1991." *Cartography and Geographic Information Science* 29, no. 3: 243–60. https://doi.org/10.1559/152304002782008431

– 2007. *Stanovlenie rubežej Rossii v Central'noj I Srednej Azii (XVIII-XIX vv.)*. Moscow, Russia: Pamâtniki istoričeskoj mysli.

Sgibnev, Wladimir. 2011. "Die blauen Kuppeln von Leninabad.: Geschichte, Charakter und aktuelle Herausforderungen einer zentralasiatischen Stadt am Beispiel des tadschikischen Chudschand." *Zentralasien-Analysen* 42: 2–11.

– 2015. "Remont: Housing Adaption as Meaningful Practice of Space Production in Post-Soviet Tajikistan. *Europa regional* 22, nos 1/2: 53–64.

Stanek, Łukasz. 2011. *Henri Lefebvre on Space: Architecture, Urban Research, and the Production of Theory*. Minneapolis: University of Minnesota Press.

Stronski, Paul. 2010. *Tashkent: Forging a Soviet City, 1930–1966*. Pittsburgh, PA: University of Pittsburgh Press.

Unverhau, Dagmar. 2006. *Kartenverfälschung als Folge übergroßer Geheimhaltung? Eine Annäherung an das Thema Einflußnahme der Staatssicherheit auf das Kartenwesen der DDR*. Münster, Germany: Lit-Verlag.

Weichhart, Peter. 1990. *Raumbezogene Identität. Bausteine zu einer Theorie räumlich-sozialer Kognition und Identifikation*. Stuttgart, Germany: Franz Steiner.

Wirth, E. 2002. *Die orientalische Stadt im islamischen Vorderasien und Nordafrika*, volume 1. Mainz, Germany: Zabern.

Conclusion

DOUGLAS YOUNG AND LISA B.W. DRUMMOND

Socialist urbanisms were and are ways of living that were intentionally shaped in societies dedicated to achieving a different social and economic organization than that of capitalist societies. Urban spaces were designed, produced, lived in, governed, and represented in ways made possible by the socialist social and economic regimes. Space was understood as crucial to the implementation of socialism and socialist urbanism. All of the contributions to this book can be understood as attempts at addressing the fundamental question: What is *socialist* about any given urban space? And as a result, what is the impact of that socialist identification on those spaces so perceived in the contemporary context? What is it about particular urban spaces that lead to their being perceived as being socialist in some way, as being representations of socialist urbanism? We identify four dimensions of space that provide a way of understanding socialist spatiality:

1. The materiality of the urban space itself;
2. Space as a shaper of a way of life;
3. The governance of space and its use; and
4. The production of space.

As the chapters separately and together indicate, socialist urbanism is socialist in so far as it exhibits a constellation of these four dimensions. Not all must be present for an urban space to be understood as socialist, but at least one of these dimensions will be. In some cases space that is materially socialist, such as housing built under socialist regimes, may no longer be governed by socialist regimes, yet has remained marked as socialist by virtue of its design, allocation, or customs of use. Those customs of use also speak to the way in which space is a shaper of everyday life, designed to embody principles about what society is and should be, how members of society should live and behave, and how society may be best and most efficiently organized. The organization of society is often a deliberate focus of the governance of space and its uses, and socialist modes of spatial governance, through mobilization, for example, or representation, may persist as long as they prove useful to the regime currently in power. Finally, the material production of space is important in terms of the technology employed, the designs considered suitable, and the interplay between material production and non-human nature in the urban environment.

The Materiality of the Urban Space Itself

The significance of the materiality of urban space to socialist and post-socialist urbanisms is clearly indicated in Hanoi and Bucharest. Drummond and Nguyen's case study (chapter 3) is of the unhappy fate of Hanoi's collective housing developments, the KTT. The present-day regime feels a need to eradicate socialist-era housing in the belief that the new developments replacing the KTT can materially represent the improved living standards and way of living brought about by the continuing implementation of market socialism. In this effort, Hanoi's experience supports Judit Bodnár's (2001) contention that societies undergoing such a transformation often place importance on eliminating the materiality of housing as a representation of an earlier socialist era. In Bodnár's study of Budapest, this took the form of a drive to privatize socialist era housing; while the housing wasn't physically erased, its collective ownership was. What is striking in the case of the Hanoi KTT is the privatization of space by residents themselves in small and large ways following the implementation of *Đổi mới* policies in the 1980s. In so doing they actively rewrote the narrative of the new and collective way of life the KTT were meant to both represent and materially shape. We see the materiality of urban space, in this case housing, being shaped by individuals and by the state, each actor operating at a different scale.

Bucharest, as chronicled by Laura Visan (chapter 8), presents a complex scenario of the materiality of space and the transition from state socialism to capitalism. While the House of the Republic, unfinished at the end of the Ceaușescu era, has been repurposed in the post-socialist era, another iconic socialist era space, the Alimentary Centre (the Hunger Circus), has been abandoned. This particular space is so profoundly imbued with its socialist goal that there is no present-day interest in repurposing it. It was produced in a style that has insufficient contemporary application to attract the interest necessary to overcome its representation of a particularly repressive socialist regime. The materiality of the space of the Alimentary Centre is saturated with meaning that is deeply unpalatable in the present era. Whereas in Hanoi the authoritarian state has dictated the material replacement of the KTT, in now-capitalist Bucharest the repurposing of the Alimentary Centre must wait for interest from property capital.

Space as a Shaper of a Way of Life

In several cities described in this book we see that space has been arranged with the intent of shaping a socialist way of life. This is particularly clear in the case studies in Stockholm, Berlin, Hanoi, and Vinh, where the durability of the original socialist intent is evident in both Stockholm and Berlin, in contrast to the Vietnamese examples (Hanoi in chapter 3 and Vinh in chapter 1). Larsson's study of households in the Stockholm satellite city, Vällingby (chapter 2), indicates that during its design and production it was clear that it was meant to represent postwar Swedish social democracy. It continues to carry that significance today, almost seventy years after its development, by way of its heritage protection status, which acts as an enforcement and reinforcement of the original intent. Larsson also found that the ideas underlying the concept of the ABC town (work, live, centre) resonate today with households in Vällingby, even as Sweden has, nationally, taken a turn to the right in many spheres including housing. These underlying ideas generate contentment with everyday life on the part of present-day residents. His interviewees represent the social diversity of residents and the benefits of living close to an array of amenities and activities, both goals of the ABC satellite city concept. In this way, Larsson's interviewees validate the social democratic and modernist ideas that drove the planning of Vällingby. The spaces of Vällingby, designed to shape a modern Swedish social democratic way of life in the 1950s, continue to shape everyday life in the present day.

Kip and Young, in chapter 9, describe the creative ambiguities associated with Alexanderplatz in Berlin East. The socialist content of this space cannot be ignored and is the basis of ongoing debates about its future. The iconic stature of Alexanderplatz in Berlin as a representation, indeed a showcase, of the GDR era is perhaps more important than the actual design. Thus its design principles are dually successful: in an iconic sense as a representation of 1960s and 70s GDR socialist urbanism, and as a social space to which people are drawn and through which they move. In the present day it continues to be a successful social space, contributing to the shaping of the everyday lives of Berliners. We ask if its continuing success is because it is socialist space, or is it simply "good space" – that is, an urban space that attracts people to it for non-ideological design reasons? While it was clearly intended to showcase GDR architecture and urban planning, perhaps its success is due to the materiality of its buildings and spaces beyond its showcase function – their arrangement, dimensions, architectural massing, scale. Or should we see in it a space that was able to be created in an era of socialism whereby its modernism distinguished it as non-capitalist? Did socialism offer an opportunity for experimentation with form because of its preoccupation with re-formatting urban space? It is these ambiguities that make Alexanderplatz a compelling case study.

The Governance of Space and Its Use

Several chapters address the governance and use of space in the socialist and post-socialist era, each focusing on a different aspect and scale of governance. In telling the housing history of one individual in socialist-era Leningrad, Borén and Gentile (chapter 4), for example, provide a case study of the experience of the socialist governance of housing in a large city. It is a story of how an individual in a highly collectivized society negotiated the allocation of housing to get the best dwelling unit possible for her household. In a society dedicated to using housing as a means of de-differentiating society, individuals in that society held a clear contradictory sense that some dwellings were qualitatively better than others. And they (some, at least, as illustrated in this chapter) had a clear sense of how to navigate the centralized housing bureaucracy to their individual benefit. All of the chapters in this book that address housing make clear that housing was an important marker of the socialist nature of urban space. While housing was intended to contribute to the de-differentiation of society, it was not experienced as uniformly equal, and in the case of Hanoi's KTT, was purposefully built and allocated unequally to reward high(er)-ranking cadre. But in the present-day post- or market socialist or postmodern era, the lived legacies of socialist-era housing are also vastly different. As Schwenkel's chapter on Vinh highlights, many of those who thought they had done well with housing initially now find themselves disadvantaged in the housing context of the present.

 With China as a case study, Carolyn Cartier (chapter 10) broadly addresses the governance of space and ways of living rather than focusing on specific city case studies or specific urban spaces. In doing so, she reveals the pliability of its socialist system. We see how a bureaucratic top-down system smoothly transitioned into a neoliberal regime focused on self-regulation. By drawing on pre-neoliberal notions of "civility," the nation state and municipal systems encourage self-regulation and thereby facilitate centralized discipline at the scale of the individual.

Ho Chi Minh City presents another example of the instrumental use of socialist-era modes of social regulation to achieve social order in a post-socialist urban society. Peyvel and Gibert's study of the transformation of leisure practices in Ho Chi Minh City (chapter 13) addresses changing ways of life and the governance of space. We see self-regulation and discipline again, achieved here through aspirational lifestyles and globalized consumption practices in the market-socialist era. We also see the enduring influence of what they call "the material and immaterial" legacies of the high socialist era, which in the case of Ho Chi Minh City lasted only a decade (1975–86). Those legacies continue to resonate more than thirty years later in ways that are supportive of maintaining social order.

The study of suburban governance at the periphery of Tirana highlights the very place-specific nature of post-socialist urbanisms. In this case, Mele and Jonas (chapter 11) describe the particular form of post-socialism produced in a country that had embraced an anti-urban socialism in which all property had been nationalized. The challenge has been to find a way of establishing property relations and an ordering of property rights in a place with new kinds of space (space that is suburban and post-socialist) and new modes of spatial and economic governance.

The case of Khujand shows how it is possible to govern through representations of space in the form of maps. Wladimir Sgibnev's focus (chapter 14) is thus on the production of representations of space, rather than of space itself, and also on ways of living – how residents of the city negotiated their way through the city with and without maps. We see how people move and remember and how erasures in the landscape are internalized by the city's residents. If we consider the broad goals of urban governance to be order, stability, and regularity, the case of Khujand demonstrates the power of spatial representations as a tool of governance in their ability to order the spatial practices of the city's residents. Sgibnev's conclusion is that there is no clear break between "a socialist and a post-socialist mode of the governance of spatial representations."

The Production of Space

Three cities in particular, Vinh, Prague, and Managua, reflect the significance of the actual production of urban space to that space being considered socialist.

The case of South City in Prague (chapter 5) makes clear the significance of the spatial production process in shaping a particular urbanism. In this case the technologies of construction (the particular form of crane that was used) negated or over-ruled the original socialist design intent. In that sense, South City represents the dictates of those technologies of construction, more so than any socialist intent to produce socialist space. Steven Logan notes that South City, in the current post-socialist era, is coming back into favour as a district – that is, it is increasingly perceived as a desirable or at least reasonable place to live. This raises the question of whether or not a design that satisfies the requirements and desires of everyday life can be both socialist and non-socialist. When is a well-facilitated everyday life ideological and when is it not ideological?

Christina Schwenkel's study of housing in Vinh (chapter 1) makes clear that in the context of Vietnam, socialist-era space only has value within the lived socialist context; outside of that context it is valueless. In fact, it turns out to be a liability today to have been assigned that space during the socialist era. Socialist-era housing was clearly a socialist-produced space and the impacts of that remain in its apparent inadaptability to the market-socialist landscape, despite, in this case, having been designed per GDR norms as self-contained units, as other KTT were not (e.g., in Hanoi). The key question is whether or not it can survive in present-day market-socialist Vietnam. This seems unlikely, as it embodies what is seen by the current regime as a representation of a failure of socialism. Ironically, current residents of socialist-era housing in Vinh have a sense that the market-socialist state needs to rescue them from high-socialist-era housing.

Managua is a case of socialist urban space, as well as pre- and post-socialist space, being co-produced by non-human nature and society. Laura Shillington (chapter 12) portrays the city as a hybrid space that reflects the relationships of different social formations (e.g., socialist, post-socialist, modern, neoliberal) with nature. And she shows us how Managua's earthquakes and lakes have shaped it as much as Nicaragua's many economic and political transformations. Thus Managua can be read as a palimpsest of urban natures that reflects the shifts they have undergone in terms of their representational significance and actual use (the shifts implemented by political and economic transformations as well as by natural events like the devastating earthquake). Inscribed in this palimpsest we can see the durability of both non-human natural processes and previous eras' ideas about organizing society and economic activity.

Complex Transitions

Two chapters in particular highlight, through a spatial lens, the complex and long-term transition from a high socialist era to varied forms of post-socialism. Phnom Penh (chapter 6), for example, as a city portrays the complex relationship between the evolution of a political and economic framework and urban spatial transformation. In the case of Phnom Penh, Gabriel Fauveaud argues that a simple temporal binary of socialist and post-socialist is overly simplistic; instead the city has to be understood as having undergone multiple temporalities from the colonial era to the present. While in many ways those political and economic temporalities have generated marked spatial transformations of the urban region, what is striking is that the centrality of the inner city has endured through the layering of new symbolic meanings. In Phnom Penh we see how space can change in terms of its meaning, while remaining largely unchanged physically.

Another particularly complex case is that of Addis Ababa (chapter 7). Jesse McClelland documents the transition in Ethiopia from a postcolonial socialism that was not urban-focused, to the present day "vanguard urbanism" of the developmental state. At work today are forces of globalization seeking to dramatically alter the urban landscape of Addis Ababa through a ramping up of redevelopment efforts, efforts that are increasingly contested. Similar to Fauveaud's contention that Phnom Penh's evolving urbanisms reflect a multiplicity of temporalities rather than a simple binary of socialist and post-socialist, McClelland depicts Addid Ababa's urbanism as reflecting a complex history of postcolonialism, conflicting socialisms, and present-day globalization.

Endnote

As noted in the introduction to this book, socialism and post-socialism are umbrella terms that cover a wide spectrum of urbanisms. In a sense, we end the book where we began by noting that there were and are many socialisms, there are a variety of post-socialisms, and there have been and continue to be complex and uneven transformation processes from past ways of organizing society and urban space to present and future ways. It is clear that the transition from socialism to what follows is an ongoing process that can stretch over a period of decades. The diverse critical reflections compiled in this book, drawing on wide-ranging cases from several continents, demonstrate the significance of the *where* and the *when* that socialist space was produced. What is also clear is that the flow of ideas about socialist and modernist space ranged across time and space, and also across political regimes in very uneven ways.

The present-day fate of socialist-era space seems not to depend on how strongly imbued that space was with socialist intent at the time of its creation as much as it does on the nature of the particular socialism that created it and of the current post-socialist situation. What is also clear is that meanings imbued in space at the time of its creation can endure for many decades and in spite of major societal shifts. In Bucharest, for example, the reviled Ceauşescu era renders some of its spaces untouchable today; in Vietnam, the market-socialist regime is determined to eliminate representations of the earlier high socialist era as a way of legitimizing the former as superior to the latter. What also seems to play a key role in determining the fate of socialist urban space is the general quality of its planning and design: 1950s Vällingby, 1960s Alexanderplatz, and 1970s South City all appear to provide attractive and socially desirable settings for urban life in the post-socialist twenty-first century. One might argue then that they represent good design in general, more so than socialist design, though on the other hand one could argue that it was the socialist orientation of the respective regimes in power at the time of their production that facilitated an urban form (one not driven by commercial concerns) that continues to be appreciated by its users today.

The contributions to this collection also demonstrate that ideas about ways of living, achieving social order, and assigning meaning to urban spaces transcend political and economic transformation, and can also outlast socialist-era space. Individuals can embody socialist-era ideas and values for decades and those values can be redeployed in the interest of post-socialist regimes and meaning associated with socialist space can be reassigned and/or reinvented. Aspects of socialist-era governance can be instrumentalized in the interest of maintaining social order in post-socialist cities. The material and symbolic production of socialist and post-socialist urbanisms is an ongoing and unpredictable process.

REFERENCE

Bodnár, Judit (2001) *Fin de Millénaire Budapest: Metamorphoses of Urban Life.* Minneapolis: University of Minnesota Press.

Contributors

Thomas Borén is Associate Professor (*docent*) in the Department of Human Geography, Stockholm University, Sweden. Since the late 1990s he has had a research interest in urban life and landscape in socialist and post-socialist cities. He defended his doctoral thesis on urban post-socialist transformation in 2005 (*Meeting-Places of Transformation: Urban Identity, Spatial Representations and Local Politics in post-Soviet St Petersburg*, published in 2009 by Ibidem-Verlag, Stuttgart, distr. Columbia University Press) and has among other things co-edited two books (2003, 2009) and one special journal issue (2007) on related topics. In other projects, often in international collaborative projects, he has researched urban policy in Sweden and other parts of Europe.

Carolyn Cartier (AB, MA, PhD University of California, Berkeley) is Professor of Human Geography and China Studies at the University of Technology Sydney and External Founding Fellow of the Centre for China in the World at the Australian National University. She is the author of *Globalizing South China* (2001, 2008, 2011) and co-editor of *The Chinese Diaspora: Place, Space, Mobility and Identity* (2003) and *Seductions of Place: Geographical Perspectives on Globalization and Touristed Landscapes* (2005). She is chief investigator of "Governing the City in China: The Territorial Imperative," an Australian Research Council Discovery Project. Her current work on cities in China focuses on establishing cities through changes to the administrative divisions and the politics and culture of urban governance.

Lisa B.W. Drummond is an associate professor of Urban Studies in the Department of Social Science at York University, Toronto. She has a PhD (Geography) from the Australian National University and a MA (Geography) from the University of British Columbia. Her work has focused on issues of urban social life in Vietnam including public space, gender, and popular culture. She has recently completed the SSHRC-funded research projects Socialist Cities in the Twenty-First Century (Berlin, Hanoi, Stockholm) with Douglas Young, and Water in the City (Hanoi and Bangkok) with Amrita Daniere. Her publications include the co-edited volumes *The Reinvention of Distinction: Modernity and the Middle Class in Urban Vientam* (with Van Nguyen-Marshall and Danièle Bélanger, 2011), *Gender Practices in Contemporary Vietnam* (with Helle Rydstrom, 2004), and *Consuming Urban Culture in Contemporary Vietnam* (with Mandy Thomas, 2003). She is working on a history of public space in Hanoi.

Gabriel Fauveaud is an assistant professor in the Department of Geography and at the Centre for Asian Studies at the University of Montreal. His current research explores the socio-spatial and socio-political aspects of the production of urban spaces in Global South cities and Southeast Asia. Fauveaud is particularly interested in real estate, transnationalism in urban planning, informal urban planning practices, and spatial justice. He has worked on Phnom Penh since 2006.

Michael Gentile is Professor of Human Geography at the University of Oslo and associate editor of *Eurasian Geography and Economics*. He has worked with various themes related to Central and Eastern Europe, including housing, socio-spatial differentiation, and urban ge-opolitics. His current regional focus is on Ukraine, and he is Principal Investigator of the Norwegian Research Council project "Ukrainian Geopolitical Fault-line Cities: Urban Identities, Geopolitics and Urban Policy."

Marie Gibert is a geographer and Assistant Professor at the University Paris Diderot (France) in the department of East Asian Studies (UFR LCAO, UMR CESSMA). Her research deals with the dynamics of public and private spaces in the production and appropriation of urban space in Vietnam. She has been conducting fieldwork in Ho Chi Minh City for more than ten years, during which time she has regularly taught urban geography at the University of Architecture and Urban Planning and the University of Social Sciences and Humanities. Her work uses a combination of process-oriented and social agency perspectives to explore the encounters between state intentions, governing practices, and everyday life during the urbanization process.

Andrew E.G. Jonas is Professor of Human Geography in the Department of Geography, Geology and Environment at the University of Hull in the UK. Andy's research interests cover various dimensions of the politics of urban and suburban development in the United States and Europe. He has written a textbook, *Urban Geography: A Critical Introduction* (Wiley-Blackwell, 2015), with Eugene McCann and Mary Thomas. His other books include *The Handbook on Spaces of Urban Politics* (Routledge, 2018), *The Urban Growth Machine: Critical Perspectives Two Decades Later* (SUNY Press, 1999), *Interrogating Alterity* (Ashgate, 2010), and *Territory, the State and Urban Politics* (Ashgate, 2012).

Markus Kip holds a PhD in sociology from York University. Working with the SSHRC-funded project The Socialist City in the Twenty-First Century, Markus co-authored the paper "Socialist Modernism at Alexanderplatz," *Europa Regional* 1–2 (2015) with his collaborators Lisa B.W. Drummond and Douglas Young. Other publications include the co-edited book *Urban Commons: Moving beyond State and Market* (2015) and the monograph *The Ends of Union Solidarity: Undocumented Labor and German Trade Unions* (2017). Markus is currently a postdoctoral researcher at the Georg-Simmel Center for Metropolitan Studies at Humboldt University, Berlin.

Bo Larsson has been working as a museum curator in Sweden since the beginning of the 1990s, mostly with contemporary research and acquisitions. He is now an employee of the City Museum of Stockholm, where he works as an editor in social sciences, as well as on scholarly assignments. Bo Larsson has a PhD in economic history and has published articles and books on museum-related topics and the everyday life of people.

Steven Logan teaches at the Institute of Communication, Culture, Information and Technology at the University of Toronto. His forthcoming book, *In the Suburbs of History* (University of Toronto Press) compares Canadian and Czech postwar suburban modernism. Steven was also the co-editor of a special issue of the journal *Public* on "Suburbs: Dwelling in Transition." The issue included a catalogue of the Leona Drive Project (2009), a site-specific art project in the Toronto suburb of Willowdale on which Steven served as researcher and co-prepared installations based on his archival research. He is currently working on infrastructure and post-suburban futures.

Jesse McClelland is a Visiting Assistant Professor in the Department of Geography at Gustavus Adolphus College (St Peter, Minnesota). In

2018, he earned a PhD in Geography from the University of Washington. Jesse's dissertation, *Planners and the Work of Renewal in Addis Ababa: Developmental State, Urbanizing Society*, enjoyed support from a Fulbright-Hays DDRA Fellowship of the US Department of Education.

Marcela Mele graduated in 2012 with a PhD in human geography from the Department of Geography, Environment and Earth Sciences at the University of Hull in the UK. She returned to Albania and is currently visiting researcher at the Gjirokastra Foundation, Albania. She previously lectured at POLIS University, the International School of Architecture and Urban Development Policies, and in the Department of Geography, University of Eqrem Çabej in Gjirokastra, Albania. Her research is interdisciplinary, incorporating urban political geography, sociology, and political economy. She is interested in how forms of market-oriented governance (a.k.a. neoliberalism) are remade through local sites and grounded practices in eastern European cities. Her PhD thesis contributes to the fields of urban transition studies and the post-socialist city.

Nguyễn Thanh Bình is Director of Urban & Rural Development Research Institute, a research unit of Vietnam Urban Planning Development Association, with architectural qualifications from the Warsaw University of Technology (Poland). He completed an MA and PhD in Cultural Heritage at Deakin University (Melbourne, Australia) and held a post-doctoral fellowship on the project Socialist Cities in the Twenty-First Century at York University (Canada). His current research explores the role of spontaneous heritage as an important human resource for cultural, social, and economic development, which also enhance urban resilience and efficiency.

Emmanuelle Peyvel is an Associate Professor of Geography at the University of West Brittany (Department of Tourism Studies) and a member of the Institute of East Asian Studies in France. Since 2005, her research has explored the development of tourism and leisure in Vietnam, especially in relation to globalization and (post)socialist nation building. Her work uses a postcolonial and intersectional approach to study unequal access to tourism resources. She regularly teaches tourism studies at Van Lang University and the National University of Economics (Ho Chi Minh City).

Christina Schwenkel is Associate Professor of Anthropology and Director of Southeast Asian Studies (SEATRiP) at the University of California, Riverside. Her research in Vietnam examines the material, technological, and cultural aftermaths of US imperial violence.

Schwenkel is author of the book *The American War in Contemporary Vietnam: Transnational Remembrance and Representation* (Indiana University Press, 2009). Her recent work on Cold War technology transfers, socialist architecture, urban decay, and postwar infrastructure has been published in *Cultural Anthropology, American Ethnologist, Critical Asian Studies*, and other journals. Schwenkel is currently co-editor-in-chief of the *Journal of Vietnamese Studies* at the University of California Press. Her book, *Building Socialism: The Afterlife of East German Design in Urban Vietnam*, is forthcoming from Duke University Press.

Wladimir Sgibnev defended his PhD degree at Humboldt University's Central Asian studies department, where he examined the social production of space in urban Tajikistan. Currently, he is senior researcher at the Leibniz Institute for Regional Geography (Leipzig), and coordinator of the institute's research group on mobilities and migration. He is working on urban processes in post-Soviet countries, with a particular focus on urban development and mobility. Recent research projects have addressed survival strategies in peripheral mining cities, informal mobilities, and a reconceptualization of public transport as public space.

Laura Shillington is a faculty member in the Department of Geoscience and the Social Science Methods Programme at John Abbott College (Montreal). She is also a research associate at the Loyola Sustainability Research Centre, Concordia University (Montreal). She holds a PhD in geography from York University (Toronto) and was a postdoctoral fellow at the Instituto de Geografáa, Universidad Nacional Autónoma de México (UNAM, Mexico City). Dr Shillington's research program broadly explores urban social-nature relations. In particular, she is interested in understanding how everyday life in urban areas, especially in mundane spaces such as the home, is embedded within multi-scalar ecological politics – from gendered human-nature relations in the household to uneven urban environmental problems and governance structures. She concentrates in particular on environmental justice and gendered and racialized experiences and knowledges of urban natures. Her research has focused on Managua, Nicaragua and, more recently, Montreal.

Laura Visan is an adjunct faculty member of the Department of Arts, Culture and Media (ACM) at the University of Toronto Scarborough and has a PhD in communication and culture from York University and Ryerson University in Toronto. She researched the process of social capital formation through civic participation and networking in the case of Romanian immigrants from Toronto. Having grown up in Romania, Laura has also written about the popular culture artefacts of the Nicolae Ceauşescu era, with a focus on the 1970s and 1980s. She has taught at the University of New Brunswick in Fredericton and at York University in Toronto.

Douglas Young is Associate Professor of Urban Studies in the Department of Social Science at York University, Toronto. He holds a B. Architecture (University of Toronto), a Postgraduate Diploma in Planning Studies (Architectural Association, London UK), and a PhD in Environmental Studies (York University). Young's current research addresses the legacies of twentieth-century socialist and modernist urbanism. His publications include *In-Between Infrastructure: Urban Connectivity in an Age of Vulnerability* (co-edited with Patricia Burke Wood, and Roger Keil; Praxis (e) Press, 2011); and *Changing Toronto: Governing Urban Neoliberalism* (co-authored with Julie-Anne Boudreau and Roger Keil; University of Toronto Press, 2009).

Index